Akanksha Bhutani

Low Temperature Co-fired Ceramics for System-in-Package Applications at 122 GHz

Karlsruher Forschungsberichte
aus dem Institut für Hochfrequenztechnik und Elektronik

Herausgeber: Prof. Dr.-Ing. Thomas Zwick

Band 93

Eine Übersicht aller bisher in dieser Schriftenreihe erschienenen Bände finden Sie am Ende des Buches.

Low Temperature Co-fired Ceramics for System-in-Package Applications at 122 GHz

by
Akanksha Bhutani

KIT Scientific Publishing

Karlsruher Institut für Technologie
Institut für Hochfrequenztechnik und Elektronik

Low Temperature Co-fired Ceramics for
System-in-Package Applications at 122 GHz

Zur Erlangung des akademischen Grades eines Doktor-Ingenieurs
von der KIT-Fakultät für Elektrotechnik und Informationstechnik des
Karlsruher Instituts für Technologie (KIT) genehmigte Dissertation

von M.Sc. Akanksha Bhutani, geb. in Dehradun, Indien

Tag der mündlichen Prüfung: 20. März 2019
Hauptreferent: Prof. Dr.-Ing. Thomas Zwick
Korreferent: Prof. Dr.-Ing. Arne Jacob

Impressum

KIT Scientific Publishing

Karlsruher Institut für Technologie (KIT)
KIT Scientific Publishing
Straße am Forum 2
D-76131 Karlsruhe

KIT Scientific Publishing is a registered trademark
of Karlsruhe Institute of Technology.
Reprint using the book cover is not allowed.

www.ksp.kit.edu

Print on Demand 2019 – Gedruckt auf FSC-zertifiziertem Papier

ISSN 1868-4696
ISBN 978-3-7315-0945-5
DOI 10.5445/KSP/1000095763

Editor's Preface

The semiconductor technologies, such as SiGe, RF-CMOS and GaAs have made tremendous progress in the last few years (e.g. the transit frequency of SiGe has reached up to 500 GHz), which has led to a high integration of circuits at frequencies in the range of millimeter-waves (mm-Waves). Simultaneously, the small wavelength at these frequencies (e.g. 2,5 mm at 122 GHz) enables either direct integration of antennas along with the active circuitry (i.e., on-chip) or within the integrated circuit package (i.e., off-chip). Therefore, the integration of complete mm-Wave radio systems in miniature-sized packages is possible. The biggest advantage in this case is that, besides the radiation transmitted and received by the antennas, no mm-Wave signal is routed in or out of the package. Consequently, the mm-Wave module can be surface-mounted on a low-cost printed circuit board (PCB) by using a standard SMD soldering process. The PCB should only be designed at very low frequencies. In the near future, a large number of mm-Wave mass products operating at frequencies above 100 GHz are expected to be available in the market. Examples include radio communication systems with very high data rates, automotive radars, applications from the areas of defence, security, space and particularly industrial automation. At mm-Wave frequencies, the performance of such a highly-integrated module is indeed extremely sensitive with respect to the material and geometry of the individual components of the package. Thus, the enormous potential of miniaturization and SMD-solderable mm-Wave systems is faced with the limited knowledge of the technological fundamentals, which are required in the development of low-cost, production-compatible packaging concepts with integrated antennas. This is exactly where the work of Dr. Akanksha Bhutani comes into play. In the framework of her activities at the Institute of Radio Frequency Engineering and Electronics (IHE) at the Karlsruhe Institute of Technology (KIT), Ms. Bhutani performed scientific research on miniaturized mm-Wave modules in LTCC technology. Her work includes investigation of concepts, optimization and realization of designs along with the active circuitry, followed by verification of the concepts. Ms. Bhutani has, for the first time, realized LTCC-based signal transitions at 122 GHz and investigated their reliability by means of a comprehensive tolerance analysis. She has successfully demonstrated two novel 122 GHz antenna concepts for integration in LTCC technology, namely, aperture-coupled patch arrays with via-fence and an electromagnetic-bandgap antenna. In addition, she has proposed a packaging concept with a partial molding-encapsulation, which enables integration of an active 122 GHz chip in the LTCC module. She has succeeded in realizing two fully-functional LTCC miniaturized radar sensors. The work of Dr. Akanksha Bhutani thus represents an important innovation to the state of the art. I am sure that her innovative concepts and remarkable demonstrators will draw much attention and find many followers worldwide. For Dr. Bhutani, with her creativity and great organizational skills, I wish her further much success in her scientific engineering career and economic endeavors.

Prof. Dr.-Ing. Thomas Zwick
- Institute Director -

Low Temperature Co-fired Ceramics for System-in-Package Applications at 122 GHz

Zur Erlangung des akademischen Grades eines

DOKTOR-INGENIEURS

von der KIT-Fakultät für
Elektrotechnik und Informationstechnik
des Karlsruher Instituts für Technologie (KIT)

genehmigte

DISSERTATION

von

M.Sc. Akanksha Bhutani
geb. in Dehradun, Indien

Tag der mündlichen Prüfung: 20. März 2019
Hauptreferent: Prof. Dr.-Ing. Thomas Zwick
Korreferent: Prof. Dr.-Ing. Arne Jacob

Kurzfassung

Diese Dissertation stellt zwei Low Temperature Co-fired Ceramic (LTCC)-basierte voll integrierte 122 GHz Radar System-in-Packages (SiPs) dar. In den beiden SiPs ist ein aktiver Radar-Chip zusammen mit LTCC-basierten passiven Komponenten (z. B. Antennen, Signalübergänge und eine Gehäusebasis für niedrige Frequenzen) integriert. Die Millimeterwellenverbindungen zwischen dem Radar-Chip und der LTCC-basierten Schaltung werden durch Drahtbonds mit einer halben Wellenlänge realisiert. Das SiP ist mit einem Epoxid-Wärmeleitklebstoff (Polytec TC 430-T) verkapselt, der eine bekannte relative Permittivität und vergleichsweise niedrige dielektrische Verluste im Frequenzbereich von 110 bis 170 GHz aufweist. Die LTCC-basierten SiPs werden jeweils auf zwei Basisband-Leiterplatten montiert, um zwei voll funktionsfähige 122 GHz Radar-Sensoren zu realisieren, deren Funktionalität durch Abstandsmessungen eines Standard-Ziels erfolgreich getestet werden konnte. Hierbei zeigt sich eine Verbesserung im Vergleich mit einem kommerziellen 122 GHz Radar-Sensor. LTCC-basierte SiPs sind zudem kompakt, kostengünstig und basieren auf Standardprozessen der Aufbau- und Verbindungstechnik.

Nach der Motivation der Arbeit wird das Ziel und die Vorgehensweise dieser Dissertation spezifiziert. Das zweite Kapitel gibt einen Überblick über den Standard LTCC-Herstellungsprozess sowie den Stand der Technik. Die Fertigungsfehler, die üblicherweise in einem LTCC-Modul auftreten können (z. B. Schrumpfung, Lagenversatz und Krümmung) werden diskutiert und analysiert. Für die vorliegende Arbeit, wird ein passendes Substratmaterial ausgewählt. Nach Berücksichtigung der Möglichkeiten sowie der Einschränkungen des LTCC-Prozesses, werden zwei LTCC-basierte SiP Konzepte vorgestellt. Im dritten Kapitel werden zwei neuartige LTCC-Signalübergänge zwischen einer Streifenleitung und einem „Grounded Coplanar Waveguide“ realisiert: ein Via-basierter Signalübergang, der über einen sehr breitbandigen Frequenzbereich (0 bis 150 GHz) funktioniert und ein elektromagnetisch gekoppelter Signalübergang. Die simulierten und gemessenen Streuparameter von den beiden Signalübergängen werden gezeigt. Außerdem werden die Prototypen der Signalübergänge durch Röntgenmikrotomographie analysiert, um den Umfang und Einfluss des LTCC-Lagenversatzes zu analysieren. Im vierten Kapitel werden mehrere 122 GHz LTCC-Antennenvarianten in Form einer „Via-fence“ und „Mushroom Electromagnetic Bandgap“ vorgestellt und daran das Prinzip der Substratwellenunterdrückung anhand elektrischer Feldsimulation und des Dispersionsdiagramms erklärt. Die Antennenvarianten unterscheiden sich durch die Speisungsmethode und die Konfiguration der Abstrahlungselemente. Die simulierten und gemessenen Antennencharakteristiken (d. h. der Reflexionskoeffizient, der Antennengewinn und die Fernfeld-Richtcharakteristik) werden jeweils dargestellt. Im fünften Kapitel werden zwei drahtgebondete Aufbauvarianten einer 122 GHz LTCC-Antenne und einem Dummy Chip, der eine 50 Ω „Grounded Coplanar Waveguide“ enthält, gezeigt. Eine der beiden Aufbauvarianten ist mit einem Epoxid-Wärmeklebstoff (Polytec TC 430-T) verkapselt. Um den Einfluss des Epoxid-Wärmeklebstoff zu bestimmen, werden die simulierten und gemessenen Charakteristiken der beiden Aufbauvarianten (d. h. mit und ohne Verkapselung) verglichen. Im sechsten Kapitel werden zwei 122 GHz LTCC-basierte Radar SiPs mit Epoxid-Verkapselung dargestellt. Die LTCC-basierten SiPs werden jeweils auf den Basisband-Leiterplatten montiert. Damit entstanden zwei einsatzbereite 122 GHz „Frequency Modulated Continuous Wave“ (FM-CW) Radar-Sensoren, deren Funktionalität durch Abstandsmessungen eines Standard-Ziels erfolgreich verifiziert wurde. Die Abstandsmessungen werden mit einem kommerziellen 122 GHz Radar-Sensor verglichen. Zum Schluss werden die wichtigsten Ergebnisse der Dissertation im siebten Kapitel zusammengefasst.

Abstract

This dissertation presents two fully-integrated 122 GHz radar System in Packages (SiPs) in Low Temperature Co-fired Ceramic (LTCC) technology. In both SiPs, an active radar chip is integrated with LTCC-based passive components including antennas, signal transitions and a low frequency package base. The millimeter-wave interconnects between the radar chip and the LTCC-based circuitry are realized through standard half-wavelength wirebonds and the SiPs are encapsulated with an epoxy molding compound, namely, Polytec TC 430-T, which has a well-defined relative permittivity and a relatively low loss tangent in the frequency range of 110 to 170 GHz. The LTCC SiPs are surface-mounted on baseband printed circuit boards, thus building two fully-functional 122 GHz radar sensors, whose functionalities are successfully tested by measuring the distance of a standard target. The performance of one of these radar sensors shows an improvement over a commercial 122 GHz radar sensor. In addition, the LTCC SiPs shown in this work are compact, low-cost and require only standard package assembly steps.

In **Chapter 1**, the primary factors, which lay the foundation for a fully-integrated millimeter-wave system are defined. Thereafter, the target and workflow of this dissertation are specified. In **Chapter 2**, the standard LTCC manufacturing process as well as state-of-the-art LTCC manufacturing techniques are presented. The manufacturing defects, commonly observed in an LTCC module, e.g., shrinkage, layer misalignment and warping, are discussed. A suitable LTCC substrate is selected for this work and two LTCC-based SiP concepts are proposed after taking the LTCC capabilities and constraints into consideration. In **Chapter 3**, a via-based stripline to grounded coplanar waveguide signal transition, operating over a broad bandwidth of DC to 150 GHz and an electromagnetic-coupling-based stripline to grounded coplanar waveguide signal transition are demonstrated for the first time in LTCC technology. Their simulated and measured scattering matrix parameters (S parameters) of both signal transitions are shown. Besides, the prototypes of both signal transitions are analyzed using X-ray microtomography, thus determining the extent and influence of the LTCC layer misalignment in each case. In **Chapter 4**, multiple variants of a via-fence and a Mushroom electromagnetic bandgap antenna, operating at 122 GHz, are presented in LTCC technology. The principle of surface wave suppression using a via-fence (i.e., a type of soft surface structure) and a Mushroom electromagnetic bandgap unit cell are expounded on the basis of the electric field and dispersion diagram simulation result, respectively. The antenna variants differ in terms of their feed and radiating element configuration. The simulated and measured antenna characteristics, i.e., reflection coefficient, realized gain and farfield radiation pattern, are shown in both cases. In **Chapter 5**, two wirebonded assemblies of a 122 GHz LTCC antenna and a dummy-chip including a 50 Ω grounded coplanar waveguide, are built with and without molding encapsulation. The simulated and measured characteristics of both assemblies are compared in order to investigate the influence of a molding encapsulation. In **Chapter 6**, two 122 GHz LTCC-based radar SiPs with molding encapsulation are presented. The LTCC SiPs are surface-mounted on baseband printed circuit boards, thus realizing two ready-to-use 122 GHz radar sensors, operating in a frequency modulated continuous wave (FMCW) mode. The functionality of these radar sensors is successfully tested by measuring the distance of a standard target and their performance is benchmarked against a commercial 122 GHz radar sensor. In **Chapter 7**, the key achievements of this dissertation are summarized.

Acknowledgement

This dissertation came into existence during my tenure as a research associate at the Institute of Radio Frequency Engineering and Electronics (IHE), Karlsruhe Institute of Technology (KIT). During this time, I have had the opportunity to work on a variety of challenging and interesting assignments. It has been a pleasure to work in a team of highly-motivated, creative and hard-working individuals and hence in a positive and cooperative workplace environment. The experiences gathered during this time have contributed significantly to my growth on both personal and professional fronts.

First and foremost, I would like to thank my supervisor and the Head of the Institute Prof. Thomas Zwick for giving me the opportunity to work on this topic and for encouraging a flexible and multifaceted problem-handling approach, which was highly-conducive to the progression and ultimately the success of this work. I would also like to sincerely thank my co-supervisor Prof. Arne Jacob for his support, constructive critique and suggestions with regards to this dissertation.

As I started working on this topic in early 2013, the beyond 100 GHz frequency range was largely an uncharted territory for the Low Temperature Co-fired Ceramic (LTCC) technology. Therefore, the LTCC components and modules, developed within the scope of this dissertation, entailed several challenges in terms of their manufacturing. I would especially thank Dr. Torsten Thelemann and Microhybrid GmbH for manufacturing the LTCC components and modules and for their effort in realizing advanced LTCC structures such as air cavities and high-resolution conducting lines. Some of the LTCC components shown in this work were analyzed using X-ray microtomography. The

X-ray analysis was highly-beneficial in investigating the extent and influence of the LTCC layer misalignment in those components. The X-ray imaging was performed by Dr. Tomáš Faragó, Dr. Thomas van de Kamp and Mr. Marcus Zuber from the Institute of Photon Science and Synchrotron Radiation (IPS) at the Test Facility and Synchrotron Radiation Source at KIT, Campus North, also known as Angströmquelle Karlsruhe (ANKA). I am truly grateful to them for investing their valuable time and effort. I also thank Dr. Nigel Smale, my mentor from the Helmholtz International Research School for Teratronics (HIRST) for his guidance. I thank Silicon Radar GmbH for providing the 122 GHz radar chips. A special thanks goes to Dr. Benjamin Göttel from Wellenzahl GmbH & Co. KG for providing the baseband module and for investing his valuable time and effort to carry out the radar measurements. I am thankful to Dr. Mario Pauli and Mr. Jörg Eisenbeis for reviewing my dissertation and for their constructive critique and suggestions on this topic. Last but not the least, I thank my parents, Brigadier Dr. Rajeev Bhutani and Deepa Bhutani for encouraging me to pursue my PhD in Germany. They have always been the pillars of my strength. I thank my sister, Srishti for her support and the youngest member of my family, Gullu, whose presence fills my heart with peace and bliss and who gives me the energy to work relentlessly.

March 2019
Akanksha Bhutani

Contents

Abbreviations and Symbols

Abbreviations

AC	Air Cavity Signal Transition
ADC	Analog-to-Digital Converter
Al	Aluminum
Al_2O_3	Alumina
Au	Gold
AUT	Antenna Under Test
BGA	Ball Grid Array
BiCMOS	Bipolar Complementary Metal Oxide Semiconductor
BN	Boron Nitride
CMOS	Complementary Metal Oxide Semiconductor
CPW	Coplanar Waveguide
CRLH	Composite Right/Left Handed
CST	Computer Simulation Technology
CTE	Coefficient of Thermal Expansion
Cu	Copper
DC	Direct Current
EBG	Electromagnetic Bandgap
E-field	Electric Field
e.g.	for example
EM	Electromagnetic
Eq.	Equation
eWLB	Embedded Wafer Level Ball Grid Array

FFT	Fast Fourier Transform
FR	Flame Retardant
FMCW	Frequency Modulated Continuous Wave
GaAs	Gallium Arsenide
GaN	Gallium Nitride
GCPW	Grounded Coplanar Waveguide
GSG	Ground-Signal-Ground
HBT	Heterojunction Bipolar Transistor
H-field	Magnetic Field
HTCC	High Temperature Co-fired Ceramic
i.e.	that is
IF	Intermediate Frequency
init	Initial
InP	Indium Phosphide
I/O	Input/Output
IQ	In-Phase and Quadrature
ISM	Industrial, Scientific and Medical
LCP	Liquid Crystal Polymer
LFPB	Low Frequency Package Base
LNA	Low Noise Amplifier
LTCC	Low Temperature Co-fired Ceramic
LWG	Laminated Waveguide
Max	Maximum
Meas.	Measurement
MEBG	Mushroom Electromagnetic Bandgap
MIMO	Multiple-Input, Multiple-Output
Min	Minimum
ML	Metal Layer
MMIC	Monolithic Microwave Integrated Circuit
mm-wave	Millimeter-Wave
Mn	Mangnese

Mo	Molybdenum
mod	Modified
MS	Microstrip
MWS	Microwave Studio
NC	No Cavity Signal Transition
OC	Open Circuit
PCB	Printed Circuit Board
Pd	Palladium
PEC	Perfect Electric Conductor
PLL	Phase Locked Loop
PMC	Perfect Magnetic Conductor
Pt	Platinum
QFN	Quad Flat No-lead
RC	Reflection Coefficient
RDL	Redistribution Layer
Rx	Receiver
SC	Short Circuit
Si	Silicon
SiGe	Silicon-Germanium
Sim.	Simulation
Si_3N_4	Silicon Nitride
SiO_2	Silicon Dioxide
SiP	System-in-Package
SL	Stripline
SMP	Surface Mount Package
SoC	System-on-Chip
SOL	Short-Open-Load
SOLT	Short-Open-Load-Through
SP	Stacked Patch
S parameter	Scattering Matrix Parameter
TC	Transmission Coefficient

TDR	Time Domain Reflectometry
TE	Transverse Electric
TL	Transmission Line
TM	Transverse Magnetic
TRL	Thru-Reflect-Line
Tx	Transmitter
W	Tungsten
WR6	Rectangular Waveguide Dimensions for 110 to 170 GHz
VCO	Voltage Controlled Oscillator
VF	Via Fence
3D	Three Dimensional

Mathematical and Physical Constants

$c_0 = 299\,792\,458\,\mathrm{m/s}$	Speed of light in free space
$\varepsilon_0 = 8.85418... \cdot 10^{-12}\,\mathrm{F/m}$	Free space permittivity
$j = \sqrt{-1}$	Pure imaginary number
$\mu_0 = 4\pi \times 10^{-7}\,\mathrm{H/m}$	Free space permeability
$\pi = 3.14159$	Archimedes' constant

Greek Symbols

α_d	Dielectric loss
β	Phase constant
β_x, β_y	Phase constant along x- and y-axis
γ	Propagation constant
Γ, X, M	Irreducible Brillouin zone
ΔR	Range resolution
ε_eff	Effective dielectric constant
ε_r	Relative permittivity
η	Antenna efficiency

θ	Surface wave incidence angle
λ_0	Free space wavelength
λ_g	Guided wavelength
μ_r	Relative permeability
ρ	Resistivity
σ	Radar cross section
ϕ	Leaky wave radiation angle
ω	Radial frequency
ω_{SE}	Series resonance frequency
ω_{SH}	Shunt resonance frequency

Latin Symbols

A_{eff}	Effective area of a target
B	Bandwidth
C	Capacitance
C_{ch}	Channel capcity
C_L	Left-handed capacitance
C_R	Right-handed capacitance
C_{via}	Via capacitance
$\overrightarrow{E}$	Electric field distribution
E_l	Longitudinal component of electric field
E_t	Transversal component of electric field
f	Frequency of operation
F	Radiation factor
f_c	Cutoff frequency
f_{IF}	Intermediate frequency of a target
f_{max}	Maximum frequency of oscillation
$FSPL$	Free space path loss
f_{start}	Start frequency of an FMCW signal
f_{stop}	Stop frequency of an FMCW signal

f_{t}	Transit frequency
H_l	Longitudinal component of magnetic field
H_t	Transversal component of magnetic field
h_{via}	Via height
k	Wavenumber
k_0	Free space wavenumber
L	Inductance
$\mathrm{L_L}$	Left-handed inductance
$\mathrm{L_R}$	Right-handed inductance
L_{via}	Via inductance
N	Number of unit cells in a periodic structure
p	Periodicity of a structure
P_{rad}	Radiated power
p_{x}, p_{x}	Periodicity of a structure along x- and y-axis
R	Line-of-sight distance
r_{ap}	Via anti-pad radius
r_{cp}	Via catchpad radius
r_{via}	Via radius
S/N	Average signal-to-noise ratio
$\tan\delta$	Loss tangent
t_{d}	Signal delay caused by a via
t_{r}	Rise time of a signal
T_{sweep}	Sweep time of an FMCW signal
X_{C}	Capacitive reactance
X_{L}	Inductive reactance
Z_l	Longitudinal component of impedance
Z_t	Transversal component of impedance
Z_{via}	Via impedance

1 Introduction

Millimeter waves (mm-waves) are electromagnetic (EM) waves with frequencies ranging from 30 GHz to 300 GHz. The corresponding free space wavelength (λ_0) in this range, varies from 10 mm to 1 mm, respectively. The idea of transmitting and receiving mm-waves is not new. In fact, the first mm-wave signaling experiments were carried out as early as in 1895, i.e., less than a decade after H. Hertz demonstrated the existence of EM waves. These experiments were performed independently by two researchers, J. C. Bose in Calcutta and P. Lebedew in Moscow, who used a λ_0 of 5 mm (60 GHz) and 6 mm (50 GHz), respectively [Eme97]. In spite of such pioneering research works, mm-wave systems were not deployed in commercial or consumer applications, up until the turn of the last century. In 1999, Mercedes Benz introduced a 77 GHz radar-based autonomous cruise control system in one of their high-end cars (Mercedes Benz S-Class) [Wen05]. Since then, the trend has started changing. A confluence of factors is driving the research and development in the mm-wave technology and it seems that the day is not far-off when mm-wave systems will play a dominant role in our everyday lives.

1.1 Motivation

Spectrum congestion

The radio wave portion of the EM spectrum, i.e., from 3 kHz to 300 GHz, is managed by regulatory agencies at both national (e.g., Federal Communications Commission in the U.S.) and international level (by International Telecommunication Union). These agencies allocate frequencies to specific applications or set of users by means of either spectrum licensing or laying down specifications for wireless devices in the unlicensed spectrum. The issue at hand is that most of the useful segments of the below 30 GHz radio spectrum are already licensed or heavily congested due to the introduction of a wide range of wireless applications, e.g., smartphones, wireless local area network, Bluetooth, ZigBee, and broadcast TV. In addition, there is an exponential increase in the number of users. On the other hand, the mm-wave spectrum has remained largely unused so far. Moreover, no specific rules (licensed or unlicensed) are defined for frequencies beyond 95 GHz, barring minor authorizations for amateur radio and industrial, scientific and medical (ISM) operations [YG18].

Advantages of larger bandwidth and higher operational frequency

A key benefit, which stems from operating at higher frequencies, is system miniaturization. In addition, all wireless applications require larger bandwidth for better performance. In a wireless communication device, a larger bandwidth provides higher data rate based on the Shannon-Hartley theorem, shown in Equation (Eq.) 1.1.

$$C_{\mathrm{ch}} = B\log_2\left(1+\frac{S}{N}\right) \tag{1.1}$$

The parameters C_{ch}, B and S/N denote the channel capacity, bandwidth, average signal-to-noise ratio, respectively. On the other hand, a radar system benefits from a larger bandwidth by achieving a better range resolution. The range resolution (ΔR) is the ability of a radar to distinguish two nearby targets. For a frequency modulated continuous wave (FMCW) radar, ΔR is obtained by using Eq. 1.2.

$$\Delta R = \frac{c_0}{2B} \tag{1.2}$$

The parameter c_0 denotes the speed of light in free space. If an FMCW radar is used to measure the range of a single target based on the intermediate frequency evaluation method (e.g., fast fourier transform), then the variance of range estimation reduces with an increase in bandwidth. If the intermediate frequency evaluation is followed by a phase evaluation, then the variance of range estimation reduces with an increase in the start frequency of the FMCW ramp. In both cases, the variance of range estimation and hence the FMCW radar accuracy is restricted by a theoretical limit named Cramér Rao lower bound [SAA+17].

Advances in semiconductor technology

The mm-wave systems for consumer and commercial applications require higher integration of active and passive components, higher speed and higher bandwidth. At the same time, these systems should have a smaller size, lower weight and lower power consumption. For such systems, the silicon-germanium (SiGe) technology is preferred over III-V compound semiconductors (e.g., gallium arsenide (GaAs), gallium nitride (GaN) and indium phosphide (InP)) due to its higher integration density, complexity, fabrication yield and cost-effectiveness [Sam18]. SiGe heterojunction bipolar transistors (HBTs) with a transit frequency (f_T) of 300 GHz and a maximum frequency of oscillation (f_{max}) of 500 GHz have been realized. As a result, monolithic microwave integrated circuits (MMICs) operating at frequencies beyond 100 GHz are available today. f_T and f_{max} are two important figures of merit of a bipolar device. Obtaining f_T and f_{max} values much higher in comparison to the operating frequency of a device is important because it provides benefits such as a lower power consumption and a lower noise figure [Cho15]. The international technology roadmap for semiconductors 2.0 has predicted that the f_T and f_{max} of SiGe HBTs will reach 650 GHz and 1100 GHz, respectively, by the year 2025 [GHM+15].

Short range applications

The range of mm-waves is limited due to the free space path loss ($FSPL$), which increases proportionally with frequency (f) and line-of-sight distance (R) squared, as shown by Eq. 1.3.

$$FSPL = \left(\frac{4\pi fR}{c_0}\right)^2 \tag{1.3}$$

The signal attenuation is further enhanced by atmospheric effects, e.g., rain, fog and moisture, and it is particularly high in certain sections of the EM spectrum, e.g., oxygen absorption at 60 GHz, water absorption at 180 GHz and 320 GHz. Therefore, mm-wave systems are mainly targeted for short range applications. As a countermeasure, at the MMIC level, the receiver sensitivity and transmit power level have been significantly improved due to the semiconductor technology advancements. In addition, at the package level, high gain and narrow beam mm-wave antennas can be designed to extend the range. Moreover, a shorter range helps in reducing the interference from nearby wireless devices.

1.2 Target

A mm-wave transceiver MMIC can be used directly for a consumer application, e.g., wireless communication link, automotive and industrial radar sensor, if it is integrated with one or more antennas and encapsulated in a package such that its functions are readily available to the end-users. In addition, the packaged MMIC should have low cost, light weight and high performance over a wide range of temperature and atmospheric conditions. For this purpose, a waveguide-based split-block package, in spite of its high performance and long-term reliability, is not suitable due to its low integration density, bulky nature and high cost. Instead, a surface mount package (SMP) is required, which can be placed directly onto the surface of a low-cost printed circuit board (PCB). Manufacturing an SMP at frequencies beyond 100 GHz entails several challenging requirements. For example, using a substrate with well-defined material characteristics and a low dielectric loss in the desired high frequency range, using a manufacturing process with a high resolution and precision. Besides, in order to ensure its commercial viability, standard manufacturing and assembly techniques should be used, which can be automated in an industrial environment. The packaging concepts are divided into two broad categories, namely, System-on-Chip (SoC) and System-in-Package (SiP). In an SoC, mm-wave antennas and other passive components are realized along with other front-end components, e.g., voltage controlled oscillator (VCO), mixer, low noise amplifier (LNA) and power amplifier, on the MMIC itself. As a result, no wirebond or flip-chip interconnect is required between the MMIC and the antennas, which in turn avoids the transmission loss and the bandwidth limitation associated with these interconnects. On the downside, antennas realized in a standard SiGe process suffer from poor bandwidth and radiation efficiency, since the antenna-to-ground distance is typically less than 15 µm, the doped Silicon (Si) substrate has a low resistivity ($\rho = 10\,\Omega$cm) and a high relative permittivity ($\varepsilon_r = 11.9$). Although high resistivity Si ($\rho < 10^5\,\Omega$cm) is available [Top14], the thickness (500 - 750 µm) and ε_r of the Si substrate cause excessive power leakage into surface waves, which can only be prevented by using special fabrication processes, e.g., through silicon vias and substrate thinning. Such processes increase the manufacturing complexity and cost [ZL09] [GWB$^+$18]. Alternatively, in an SiP, mm-wave antennas and other passive components are realized on a separate dielectric substrate, hence wirebond or flip-chip interconnects are required between the MMIC and the transmit (Tx) and receive (Rx) antennas. The conceptual block diagram of an SiP is shown in Fig. 1.1. At mm-wave frequencies beyond 100 GHz, both wirebond and flip-chip interconnects introduce parasitic effects, which should be compensated in order to achieve sufficient reflection coefficient (RC) and transmission coefficient (TC) values in the desired frequency range. On the positive side, an SiP provides flexibility in selecting a suitable substrate and technology for realizing antennas with a high bandwidth-efficiency product. This work focuses on the implementation of an SiP operating at a frequency of 122 GHz and the technology used for this purpose is Low Temperature Co-fired Ceramic (LTCC). Among the various packaging technologies available today, the LTCC technology is particularly attractive, because all SiP components listed below can be integrated in a single, compact, three-dimensional (3D) LTCC module.

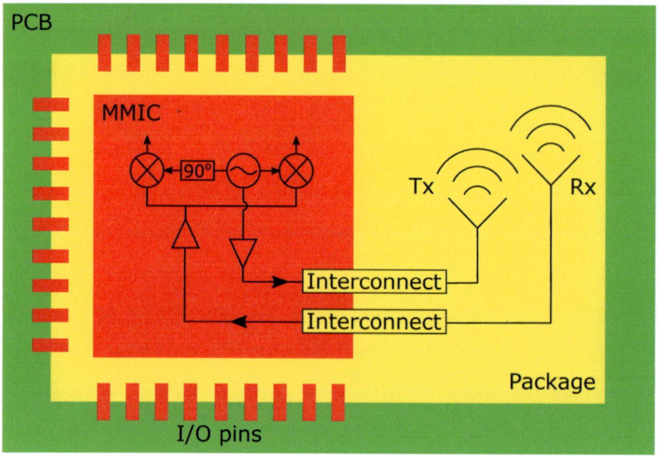

Figure 1.1: Conceptual block diagram of an SiP

1 *Off-chip circuit implementation*: LTCC is a multilayer substrate technology. It can be used to integrate a variety of features, including conducting lines, vias, air cavities and fluidic channels in a compact 3D module. Therefore, mm-wave antenna arrays with a non-planar feeding mechanism, such as aperture-coupling, can be realized in LTCC. A stripline (SL) or a substrate integrated waveguide could be used to feed an antenna array. Other passive components, e.g., signal transition and coupler, can also be implemented in the same 3D module.

2 *Housing, thermal management and encapsulation of an MMIC*: These three functions can be achieved single-handedly by using the LTCC technology. First, an MMIC can be placed in an LTCC cavity and it can be connected to the off-chip LTCC circuit by using either wirebond or flip-chip interconnects. Second, thermal vias and/or fluidic channels can be realized in an LTCC module, thus dissipating the heat generated by the MMIC. Third, the MMIC can be encapsulated by means of either a hermetic lid brazed to the LTCC module or molding encapsulation.

3 *Interface between an MMIC and a PCB*: A ball grid array (BGA) can be realized on the bottom of an LTCC module, thus acting as an interface between the MMIC and a low-cost PCB. An LTCC module is well suited for the reflow soldering process, since it can withstand very high temperatures.

There are also some challenges involved in working with LTCC. These challenges are mainly related to the deviations, which occur during the LTCC manufacturing process, e.g., substrate shrinkage, layer misalignment and warpage. At mm-wave frequencies beyond 100 GHz, these deviations could significantly alter the electrical characteristics of the LTCC components. Therefore, the influence of these deviations on the electrical characteristics of the LTCC components has been carefully investigated, both at design level through full-wave EM simulations and post manufacturing by means of X-ray analysis.

1.3 Workflow

The organization of this dissertation is shown with the help of a flow chart in Fig. 1.2. Chapter 2 begins with a description of the standard LTCC manufacturing process, followed by an overview of the state-of-the-art manufacturing techniques. Next, the manufacturing deviations and constraints are discussed. For the task at hand, a suitable LTCC substrate is selected and two SiP concepts are presented, namely, a vertical-configuration SiP and a horizontal-configuration SiP. The pros and cons of their implementation at frequencies above 100 GHz are highlighted.

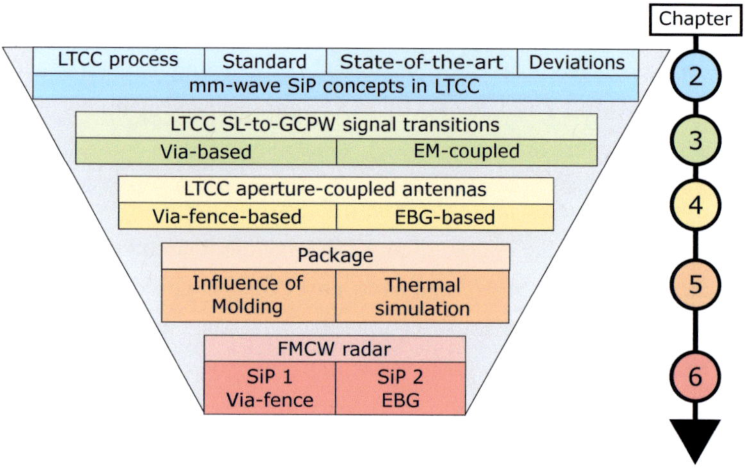

Figure 1.2: Flow chart showing the organization of this dissertation

In Chapter 3, a via-based and an EM-coupling-based signal transition between a grounded coplanar waveguide (GCPW) and an SL are presented. The prototypes of both signal transitions are investigated by means of X-ray microtomography and hence the deviation of buried conductors and vias caused by the LTCC manufacturing process are analyzed. Based on the X-ray analysis, the initial simulation model is modified in each case and the initial and modified scattering matrix parameters (S parameters) are compared with the measured S parameters.

In Chapter 4, variants of an aperture-coupled via-fence antenna and an aperture-coupled electromagnetic bandgap (EBG) antenna are presented. These antenna variants differ in the following two aspects. First, feedline: the antennas employ either a microstrip (MS)-to-GCPW transition or an SL-to-GCPW transition. Second, number of radiating elements: for the via-fence antenna, a higher antenna gain is achieved by using an array of 2×2 elements and for the EBG antenna, a higher antenna gain is achieved by using a larger number of EBG unit cells.

In Chapter 5, a proof of concept is built by wirebonding a dummy chip including a 50 Ω GCPW line to an aperture-coupled via-fence stacked-patch antenna. The antenna is fed by an SL-to-GCPW transition. The dummy chip and antenna assembly is encapsulated by using an epoxy molding compound, namely, Polytec TC 430-T. The dielectric properties of this molding compound have been characterized in the frequency range of 110 to 170 GHz and it has a relatively low dielectric loss in this frequency range. The influence of the molding encapsulation on the bandwidth and radiation characteristics of the antenna are analyzed. In addition, the thermal simulation result of the proposed horizontal-configuration LTCC SiP is presented.

In Chapter 6, two LTCC-based SiPs are presented. In the first SiP, the off-chip circuit consists of two 2×2 aperture-coupled via-fence antenna arrays, fed by two SL-to-GCPW transitions. In the second SiP, the off-chip circuit consists of two 6×6 EBG antennas, fed by two SL-to-GCPW transitions. Both SiPs house a 122 GHz SiGe bistatic radar transceiver MMIC. Each of these SiPs is mounted on a PCB and used to measure the distance of a standard target (i.e., a trihedral corner reflector). The performance of both SiPs is benchmarked by comparing the measurement results to that of a commercially available 122 GHz radar frontend.

2 LTCC-based SiP Concepts for mm-Wave Applications

LTCC is a multilayer ceramic substrate technology, in which metals and metal alloys with high electrical conductivity, e.g., silver (Ag), gold (Au), Ag - palladium (Pd), Ag - platinum (Pt) and Au-Pt, are co-fired with ceramic substrates at a temperature of 800 to 1000 °C. The LTCC technology is primarily a derivative of the High Temperature Co-fired Ceramic (HTCC) technology, in which alumina (Al_2O_3) substrate is co-fired with conductor materials, e.g., molybdenum (Mo), tungsten (W) and manganese (Mn), at a much higher temperature of 1600 to 1800 °C. The melting point of the conductor materials used in LTCC is much lower than in HTCC (e.g., the melting point of Au is 1063 °C, whereas the melting point of W is 3410 °C), which forces the LTCC co-firing temperature to be kept lower than the melting point of the metal used. In addition, amorphous glass, crystallized glass, modifier oxides etc. are added in specific amounts to the LTCC substrate, thus ensuring a high sintering density at a low firing temperature [Ima05]. The main characteristics of the LTCC technology, which establish its suitability for designing an SiP for mm-wave applications are as follows.

1 *Miniaturization*: Multiple substrate layers, typically ranging between 3 and 40 layers, integrated with transmission lines (TLs), vias, air cavities and fluidic channels, can be stacked vertically to implement a variety of passive components in a compact 3D module [Gmb]. In addition, resistive and dielectric inks can also be co-fired in an LTCC module. Therefore, monolithic integration of buried resistors, capacitors and inductors within a limited range of values is feasible [Bec09]. Conse-

quently, LTCC has a strong edge over single- and double-sided ceramic substrate technologies in terms of integration density and miniaturization of SiPs.

2 *Low dielectric loss substrate*: At mm-wave frequencies, the dielectric loss ($\propto f$) dominates over the conductor loss ($\propto\sqrt{f}$). The dielectric loss of a substrate (α_d) in $\mathrm{dB\,m^{-1}}$ is given by Eq. 2.1 [Ima05]. The parameters ε_r and $\tan\delta$ denote the relative permittivity and loss tangent of the dielectric substrate, respectively. There are a few LTCC substrates available with relatively low values of $\tan\delta$ at mm-wave frequencies.

$$\alpha_\mathrm{d} = 2.73\frac{f}{c_0}\sqrt{\varepsilon_r}\tan\delta \qquad (2.1)$$

3 *Thermal stability*: The assembly process of an SiP involves steps such as mounting active and passive components on a package, soldering the SiP on a baseband PCB, which lead to thermal stress. The coefficient of thermal expansion (CTE) of LTCC substrates is relatively low over a very wide temperature range, e.g., CTE of DuPont™ GreenTape™ 9K7 (DuPont 9K7) is 4.4 ppm/K between 23 and 300 °C [Mat09]. As a result, LTCC modules are highly stable, even when subjected to thermal shocks. Moreover, the CTE of most LTCC substrates is closely matched to the CTE of semiconductors, e.g., SiGe and GaAs [Mat18]. Therefore, SiGe and GaAs MMICs can be reliably connected to LTCC packages by using wirebond or flip-chip interconnects. Although the thermal conductivity of LTCC substrates is relatively low, e.g., 4.6 W/mK for DuPont 9K7 [Mat09], the heat generated by an active MMIC can be efficiently dissipated by using thermal vias (in a standard LTCC package) or fluidic channels (in an advanced LTCC package).

4 *Wirebond adhesion and solder wetability*: The connections between an LTCC module and an MMIC are usually realized using wirebonds. In

applications, where a high connection reliability is desired, it is possible to use Au metallization with an LTCC module, which facilitates wirebonding. Additionally, the solder wetability of Au and its alloys is excellent, which helps in forming solder joints between an LTCC module and a baseband PCB.

In Table 2.1, a comparison of the main properties of the following multilayer substrates is shown: a) Rogers Ultralam 3850HT - a high-temperature liquid crystal polymer (LCP) laminate b) 92% Al_2O_3 substrate, commonly used for multilayer HTCC packages c) DuPont 9K7 LTCC material - a lead free, low-loss glass-ceramic composite material. It is observed that in terms of dielectric loss and thermal stability, DuPont 9K7 (LTCC) is superior to both Rogers 3850HT (LCP) and Al_2O_3 (HTCC). In terms of thermal conductivity, Al_2O_3 shows the best performance, followed by DuPont 9K7 and Rogers 3850HT. Finally, among the three substrates, only LTCC provides Au metallization (without requiring any special manufacturing techniques), which makes it cost-effective and preferable for forming reliable wirebond and flip-chip interconnects.

2.1 Selection of an LTCC Substrate

The electrical, thermal and mechanical properties of various commercial LTCC substrates are shown in Table 2.2 for comparison. The following observations are made on the basis of this table. First, there are three LTCC substrates available, whose dielectric properties are characterized up to and beyond 100 GHz, namely, DuPont 9K7, Ferro A6M and Heraeus CT800. Their $\tan\delta$ values lie between 0.001 and 0.002 at around 100 GHz. Among these three substrates, Heraeus CT800 is a constrained sintering (described in section 2.4) LTCC tape [IKT]. Second, the lowest value of CTE is observed for Kyocera GL570 (3.4 ppm/K), followed by DuPont 9K7 (4.4 ppm/K). Third, the highest thermal conductivity is shown by DuPont 9K7 (4.6 W/mK), followed by Heraeus CT800, Heraeus CT708 and Kyocera GL330 (4.3 W/mK). Fourth, the highest flexure strength is provided by Kyocera GL330 (400 MPa), followed by DuPont 951 (320 MPa). The electrical, thermal and mechanical properties of DuPont 9K7 are found to be optimum and hence it is used for designing the mm-wave components in this work. DuPont 9K7 has been characterized up to frequencies as high as 750 GHz [GO15].

Substrate	Electrical		Thermal		Metal
	ε_r	$\tan\delta$	CTE	Conductivity	
	at 10 GHz, 23 °C		ppm/K	W/mK	
Rogers 3850HT (LCP) [Cor15b]	3.14	0.002	18 (30 - 150 °C)	0.2	Cu
92% Alumina (HTCC) [Cer]	9.2	0.003	6.57 (25 - 300 °C)	20.3	Mo, W, Mn
DuPont 9K7 (LTCC) [Mat09]	7.1	0.001	4.4 (23 - 300 °C)	4.6	Au, Ag, Cu, Ni, Au-Pt etc.

Table 2.1: A Comparison of the properties of commercial multilayer substrates

Substrate	Electrical			Thermal		Flexure strength
	f	ε_r	$\tan\delta$	CTE	Conduc-tivity**	
	at f in GHz, 23 °C			ppm/K	W/mK	MPa
DuPont 951 [Mat09]	10	7.8	0.014	5.8	3.3	320
DuPont 9K7 [GO15]	60	7.05	0.0014	4.4	4.6	230
DuPont 9K7 [GO15]	500	7.2	0.0048	4.4	4.6	230
ESL 41020 [Eleb]	10	7.8	0.011	7.4	2.5-3.0	-
Ferro A6M [Cor15a]	95	6.1	0.0016	7.0	2	170
Heraeus CT800 [IKT]	100	5.2	0.002	5.1	4.3	-
Heraeus CT708 [IKT]	2.5	6.4	0.003	9.3	4.3	-
Heraeus HL2000 [Elea]	2.5	7.3	0.0026	6.1	-	>200
Kyocera GL330 [Kyo]	2	7.7	0.0005	8.2	4.3	400
Kyocera GL570 [Kyo]	2	5.7	0.0007	3.4	2.8	200
Kyocera GL580 [Kyo]	2	6.1	0.0016	10.4	2.0	270
Kyocera GL771 [Kyo]	2	5.2	0.0036	12.3	2.0	170
Kyocera GL773 [Kyo]	2	5.8	0.0023	11.7	1.9	280
Amkor GCS71	10	7.1	0.005	5.5	3.2	280

Table 2.2: A Comparison of the electrical, thermal and mechanical properties of commercial LTCC substrates (**Thermal conductivity from 23 to 300 °C)

The key advantages of DuPont 9K7 are listed below [Mat09].

a A highly-stable dielectric constant and a low dielectric loss up to and beyond 100 GHz

b Compatibility with a wide variety of metal pastes, which are suitable for wirebonding, flip-chip and soldering

c Ability to withstand harsh environmental conditions

d Hermetic packaging at a relatively low temperature (< 500 °C)

e High volume manufacturing at lower total system cost

The important physical properties of DuPont 9K7 LTCC substrate post firing are given in Table 2.3 [Mat09]. In this work, all mm-wave components and modules are manufactured using DuPont 9K7PV substrate.

Physical properties	DuPont 9K7
Average thickness (µm)	106.7 (9K7PV) 213.4 (9K7PX)
X,Y shrinkage (%)	9.1 ± 0.3
Z shrinkage (%)	11.8 ± 0.5
Surface roughness (µm)	0.52

Table 2.3: Important physical properties of the selected LTCC substrate - DuPont 9K7 post firing

2.2 Manufacturing Aspects of LTCC Technology

2.2.1 Standard LTCC Manufacturing Process

In order to design robust mm-wave components in LTCC technology, it is important to have a basic understanding of the various stages involved in the

LTCC manufacturing process and the macro- and micro-level defects that can arise at each stage. The standard LTCC manufacturing process is divided into six stages, out of which co-firing or sintering is the last stage. Before co-firing, an LTCC substrate layer is commonly known as 'Green Tape'. The manufacturing stages are briefly described below. An in-depth explanation of the LTCC manufacturing process is given in [Ima05]. A block diagram of the manufacturing process is shown in Fig. 2.1.

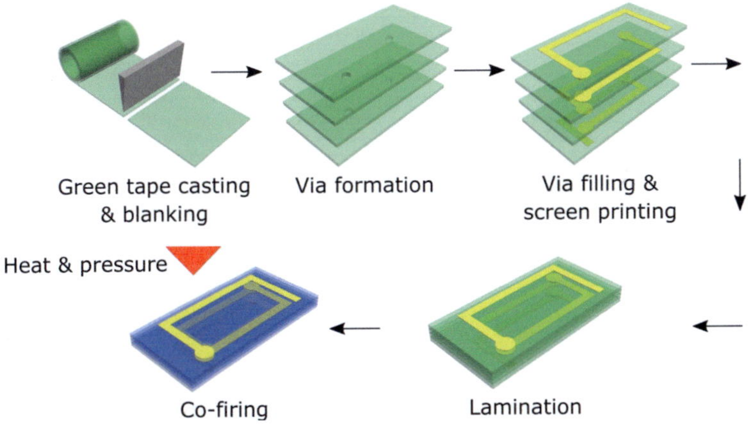

Figure 2.1: Standard LTCC manufacturing process

1 *Preparation of ceramic slurry*: The raw materials including inorganic materials (e.g., ceramic and glass powder) and a variety of organic materials (e.g., binders, plasticizers, dispersing agents and solvents) are mixed in a specific ratio to prepare a slurry. Although the organic materials are added in very small amounts and they are completely eliminated during the co-firing process (in the ideal scenario), they play an important role in determining the characteristics of the final LTCC module. For example, the binder prevents formation of cracks in the green tape during the casting process and it also ensures that no air gaps are formed between the green tapes during the lamination process.

2 *Green tape casting and blanking*: Casting is the process of molding the homogenous slurry of raw materials into two-dimensional green tapes. The fundamental technique used in casting machines is known as the doctor blade method. In this method, the slurry is fed into a tank, which has a small gap at its bottom (i.e., the doctor blade). As the volume of the slurry in the tank increases, it exerts a downward pressure, which forces a thin film of the slurry to flow out through the doctor blade. The whole setup is mounted on a plastic carrier film, which moves at a constant speed by using a conveyor belt. The slurry ejected from the blade, dispensed on the carrier film, is dried in hot air and thus transformed into a green tape. The following green tape characteristics are controlled during the casting process. First, the green tape thickness is controlled by adjusting external factors, e.g., blade gap, carrier film speed and slurry height in the tank. Second, the drying should be slow and homogenous, otherwise it leads to micro- or macro-level defects, e.g., formation of cracks and blisters and bending of the green tape. The green tape is subsequently cut into panels.

3 *Via Formation*: Conventionally, via holes of diameter greater than 100 µm are punched or drilled in the green tapes. The green tape should neither be too brittle nor too soft, as both cases could lead to via defects. In the former case, the via punch causes undesired chipping of the green tape and in the latter case, it leads to adhesion of scrap material on the via hole edge. This scrap material may eventually block the via holes, resulting in conductive defects. Due to the recent developments in the LTCC manufacturing technology, fine via holes with diameter less than 100 µm, i.e., microvias, have become feasible.

4 *Via filling and screen printing*: The via holes formed in the green tapes, using a punch, drill or laser, are filled with conducting material. The via holes should not be packed too densely with the conducting paste because the green tape undergoes shrinkage during the co-firing process,

which decreases the volume of the via holes. This leads to development of stress and hence via distortion. On the other hand, if the conductor packing density in the via holes is insufficient, it could result in an open circuit. After via filling, the conducting patterns are formed on the green tapes by using a screen printing process. In this process, a mask is applied on the green tape and the conductive paste is pressed through the mask opening onto the green tape.

5 *Lamination*: The green tapes with metal patterns and vias are vertically aligned to ensure proper connection of the 3D circuit. The alignment is done in the following manner: Each green tape is placed on an XY-theta alignment stage. The green tape has four markers, i.e., one at each corner, which are matched with optical axes using cameras. After alignment, the green tapes are pressed together in a mold and simultaneously subjected to heat ($\approx 80\,^{\circ}$C) and pressure (≈ 30 MPa). As a result, the tapes are combined into a single unit. During this step, shrinkage along x-, y- and z-axis may take place.

6 *Co-firing*: During the co-firing or sintering process, the laminated body is subjected to a high temperature (800 to 1000 °C) and pressure, which causes coalescing of particles. The complete mass including ceramic and metal wiring undergoes shrinkage. There are two key requirements of the co-firing process. First, the mismatch in the shrinkage rate of the ceramic and metal should be minimized. Second, the variation in the shrinkage rate of the whole mass should also be minimized. These requirements depend on factors such as ceramic/glass composition, co-firing temperature profile, co-firing setup and the volume percentage of metal in the module.

2.2.2 State-of-The-Art Manufacturing Techniques for High-Resolution LTCC Structures

Modules operating at mm-wave frequencies have two basic requirements. First, the selection of a dielectric substrate with a well-defined dielectric constant and a low loss tangent in the desired frequency range. This aspect is dealt with in the next section. Second, the ability to manufacture TLs with narrow line/space width and microvias. In addition, the manufacturing tolerances of these high-resolution structures should be sufficiently low. The minimum line/space width achievable with the conventional screen printing process is typically 100 µm. In contrast, the minimum line/space width achievable with thin film technology, using single- or double-sided ceramic substrate (e.g., Al_2O_3) is typically less than 30 µm. Over the last decade, a couple of novel manufacturing techniques have been developed to produce high-resolution structures in the LTCC technology. The highlights of these state-of-the-art manufacturing techniques are listed below.

1 *Modification of the standard screen printing process*: The standard screen printing process can be modified by using one of the following ways: a) Special calendared screens with a high mesh count and fine wire diameter b) Photo-sensitive pastes c) Photo-etching technology. These technologies can be applied before or after the sintering process, depending on the metal paste used. The minimum line/space width achievable by using these methods is 30 µm. The best case manufacturing tolerance is $\approx 5\%$ for a target line width of 100 µm and $\approx 12\%$ for a target space width of 50 µm [MPT+06].

2 *Thin film deposition*: This technique is used to realize high-resolution structures on an LTCC substrate post firing, hence it is only applicable to the top and bottom substrate layers. The maximum resolution achievable with this technique is 10 µm. The manufacturing tolerance

is $\approx 5\%$ for a target line and space width of 100 µm and 50 µm, respectively [MPT+06].

3 *Laser ablation*: The manufacturing processes mentioned above entail massive manufacturing effort and are incompatible with the standard LTCC process [GSW+17]. These issues are resolved by using a laser ablation method. This technique has been extensively used in this work, an example of which is shown in Fig. 2.2. The TL shown in this figure has a target line and space width of 60 µm and 40 µm, respectively (as per design). The manufactured line and space width of the TL are approximately 55 µm and 48 µm, respectively, as measured by a high-quality microscope. In this method, a solid metal layer is first screen printed on the LTCC substrate. Thereafter, a laser beam is impinged on specific areas of the metal surface to create high-resolution structures. The laser beam specifications, e.g., average laser power, pulse duration and pulse repetition rate, are precisely controlled so that only the desired portion of metal or substrate layer is ablated by the laser beam, without generating excessive heat [GSW+17]. The laser ablation technique has the following advantages.

   a Laser ablation of screen-printed lines can either be done before or after the co-firing process. Therefore, this technique can be used for producing high-resolution structures on the outermost as well as on a buried substrate layer.

   b The maximum achievable accuracy for line/space width is better than ± 1 µm [GSW+17].

   c It can also be used for drilling microvias with a diameter of 50 µm and air cavities in an LTCC module.

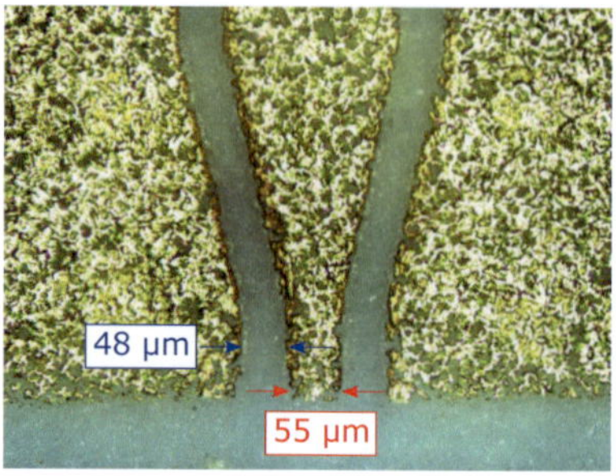

Figure 2.2: High-resolution structure on an LTCC substrate

2.2.3 Pitfalls of LTCC

The key manufacturing constraints of the LTCC technology, which should be taken into consideration while designing a mm-wave SiP are as follows:

1 *Shrinkage*: During an unconstrained co-firing process, the substrate and metal layers undergo shrinkage along the x-, y- and z-axis. In contrast, a constrained co-firing process induces shrinkage only along the z-axis. The latter is time consuming and expensive, hence it is not used in this work. The shrinkage rate depends on various factors, such as material composition, firing temperature profile and firing setup. Therefore, the substrate height as well as the metal pattern post firing show finite tolerances. For DuPont 9K7PV substrate, the unfired substrate thickness is 127±9 µm and the substrate shrinkage along the z-axis is 11.8±0.5% [Mat09]. Based on these values, it is estimated that the fired substrate thickness can vary between 103.5 and 120.6 µm.

2 *Warping*: During the co-firing process, warping of an LTCC module usually occurs due to the mismatch in the shrinkage rate of the substrate and the metal paste. Moreover, an uneven co-firing process, which occurs due to factors, such as asymmetric temperatures in the furnace and the reaction between the LTCC module and the setter (i.e., the platform

on which an LTCC module is placed during the co-firing process) can also lead to warping. Warping of an LTCC module is undesirable, since it distorts the circuit embedded in a 3D LTCC module. As a result, it becomes difficult (or impossible) to contact the module with a probe for S parameter measurement. In addition, it becomes difficult (or impossible) to achieve a surface-mount assembly [Low18]. In order to avoid warping, the following two constraints should be followed. First, the number of substrate layers should not be too few (i.e., the LTCC module should not be too thin) [Ltd]. The LTCC manufacturer selected for this work recommends at least 3 or 4 substrate layers of DuPont 9K7PV. Second, the metal volume percentage in an LTCC module should not exceed a certain limit. In some cases, the metal volume percentage is reduced by using a meshed ground plane instead of a solid ground plane [Ltd], an example of which is shown in section 4.1.1.

3 *Layer Misalignment*: During the lamination process, a misalignment of substrate layers may take place. The layer-to-layer misalignment lies typically in the range of 20 to 50 µm. Therefore, in the worst case scenario, an LTCC module with 10 substrate layers can have a misalignment of 450 µm between the top and bottom substrate layers, i.e., approximately $\lambda_g/2$ at 122 GHz for DuPont 9K7 substrate (The parameter λ_g denotes the guided wavelength). Such a large misalignment between the layers can severely degrade or even nullify the electrical performance of an LTCC module. Therefore, the usage of too many substrate layers has been avoided in this work. The mm-wave LTCC components shown in this work are implemented with a maximum of 5 substrate layers.

4 *Fabrication of air cavities*: Air cavities are often required in an SiP for mechanical as well as electrical purposes, e.g., to place an MMIC and to enhance the radiation efficiency of an antenna. In principle, there are two types of air cavities, which can be fabricated in an LTCC module,

namely, a surface and an embedded air cavity. The main challenge involved in fabricating an air cavity is to optimize the lamination pressure for a given substrate thickness and cavity depth [TTH02]. In case the lamination pressure is too high, it leads to deformation of the cavity (i.e., sagging) as well as the substrate. The sagging of a cavity can be avoided by using a sacrificial volume material, e.g., carbon, which increases the manufacturing complexity and effort [Mal12]. In this work, only surface cavities have been used.

2.3 mm-Wave Surface-Mount SiP Concepts

In this section, two LTCC-based SiP concepts are described, which form the basis of this work. The SiP concepts differ in terms of the alignment of a transceiver MMIC and mm-wave off-chip antennas, which are used for transmitting and receiving the signal. As a result, the following SiP design requirements are impacted. First, the number of substrate layers. Second, the signal transition interconnecting the MMIC and the TL feeding the antennas. Third, thermal management of the MMIC. The key aspects of these SiP concepts and the challenges involved in implementing these concepts at mm-wave frequencies beyond 100 GHz are discussed. Later in this section, an overview of mm-wave SiPs realized in other contemporary technologies is given, followed by a list of the most significant LTCC-based mm-wave SiPs demonstrated in the recent years.

2.3.1 Vertical-Configuration SiP Concept in LTCC

A vertical-configuration SiP concept is shown in Fig. 2.3. In this concept, the MMIC is placed face down in a multistep surface cavity, at the bottom of an LTCC package. The antenna and MMIC are aligned vertically within the package, thus reducing the package footprint. An aperture-coupled antenna is used, which is fed by a buried-conductor TL, e.g., an SL or a laminated waveguide (LWG). Consequently, a signal transition is required between the buried-conductor TL and an open-conductor TL (e.g., GCPW), which in turn forms a mm-wave interconnect between the LTCC package and the MMIC. The signal transition is based on EM coupling of the buried- and open-conductor TLs. Alternatively, if a via-based signal transition is used, the signal via interconnecting the two TLs will have a height of approximately 320 μm (i.e., the signal via extends over 3 layers of DuPont 9K7PV substrate). This in turn will lead to large parasitic effects at mm-wave frequencies beyond 100 GHz. The parasitic effects can be mitigated by means of an embedded air cavity, placed directly above the signal via, but realizing an embedded air cavity in an LTCC module is highly susceptible to manufacturing tolerances at present. A via-based and an EM-coupled SL-to-GCPW signal transition are discussed in detail, in section 3.1 and 3.2, respectively. The mm-wave interconnect between the LTCC package and the MMIC consists of wirebonds, which also result in parasitic effects due to the high frequencies involved. These parasitic effects can be compensated in one of the following ways. First, by integrating a TL-based compensation network in the LTCC package. Such a network usually consists of a high impedance TL stub, which requires a narrow line width of around 20 μm [BRG+13]. As mentioned in section 2.2, such narrow line widths have been demonstrated only on basic test structures in the LTCC technology and are therefore avoided in this work. Second, wirebonds of length, $\lambda_0/2$ (if the medium is air, i.e., the package is encapsulated with a hermetic lid as shown in Fig. 2.3), compensate the parasitic effects due to their self-matching property. At 122 GHz, the value of $\lambda_0/2$ is 1.23 mm, which can be easily manufactured using the standard wirebond technology [Bee13]. The mm-wave signals are confined within the LTCC package and the low frequency signals, i.e., direct current (DC) and baseband signals, required for operating the MMIC, e.g., voltage supply and intermediate frequency (IF) signal, are routed from the MMIC to a PCB (or vice versa) via the LTCC package. The low frequency interconnects between the MMIC and the LTCC package are realized through standard wirebonds (without any compensation). In addition, the low frequency interconnects between the LTCC package and a PCB are realized by means of a BGA. The MMIC is encapsulated by using either a hermetic lid, as shown in Fig. 2.3 or a molding compound. In both cases, the encapsulation should have practically no influence on the antenna radiation characteristics. The manufacturing and assembly effort required for implementing the vertical-configuration SiP is assessed on the basis of the following points:

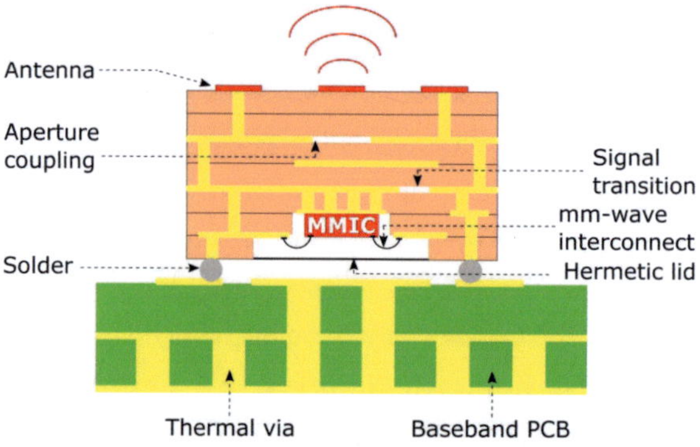

Figure 2.3: Vertical-configuration SiP concept

1 The vertical-configuration SiP requires at least 7 substrate layers for its implementation. Therefore, the layer-to-layer misalignment should be strictly controlled in order to ensure a high electrical performance.

2 The heat generated by the active MMIC is dissipated by attaching the MMIC to a metal pad, backed by thermal vias. The thermal management in this case could be challenging due to the limited package footprint and the long thermal path. In principle, it is possible to integrate fluidic channels in an LTCC substrate [TTH02], which serve as an active heat sink for the MMIC, but this approach would increase the number of substrate layers as well as the manufacturing complexity involved.

3 The assembly effort of a vertical-configuration SiP is coarsely divided into the following steps: a) Placing and gluing the MMIC in a multistep

cavity at the bottom of the LTCC package. A thermally conductive adhesive should be used for this purpose. b) Forming the mm-wave (i.e., $\lambda_0/2$), DC and baseband wirebond interconnects between the LTCC package and the MMIC c) Brazing a hermetic lid to the LTCC package d) Soldering the vertical-configuration SiP to a baseband PCB.

2.3.2 Horizontal-Configuration SiP Concept in LTCC

A horizontal-configuration SiP concept is shown in Fig. 2.4 [10]. The off-chip circuit of the SiP is divided into two parts, namely, a mm-wave module and a low frequency package base (LFPB). The mm-wave module consists of an antenna and a signal transition, which is required for interconnecting the antenna and the MMIC. The LFPB consists of the following components. First, a lead frame on the periphery, which interconnects the SiP to a baseband PCB. Second, thermal vias connected by two large metal pads, which help in dissipating the heat generated by the MMIC. Third, a surface cavity, which facilitates SiP encapsulation either by means of a lid or a molding compound. The combination of the mm-wave module and the LFPB produces a multi-step cavity structure, in which the MMIC is placed face up. The MMIC and the antenna are aligned horizontally in this concept. This results in a larger package footprint than the vertical-configuration SiP. On the positive side, the horizontal-configuration SiP has several advantages in terms of its manufacturing and assembly, which promises a robust and flexible package design at mm-wave frequencies beyond 100 GHz. Similar to the vertical-configuration SiP, the horizontal-configuration SiP also uses an aperture-coupled antenna, fed by a buried-conductor TL, but unlike the vertical-configuration SiP, the signal transition between the buried- and open-conductor TL is achieved by using a signal via. In this case, the signal via has a height of approximately 105 μm (i.e., the signal via passes through only one layer of DuPont 9K7PV substrate). Therefore, the signal discontinuity arising as a result of the signal via is much lower than in the vertical-configuration SiP. Additionally, a sur-

face air cavity can be realized below the signal via, in order to mitigate the parasitic effects and thus improve the RC of the signal transition in the desired frequency range. This topic is covered in detail in section 3.1. The MMIC and antenna are interconnected through self-matched $\lambda_g/2$ wirebonds and the horizontal-configuration SiP, barring the radiating elements on the top of the antenna, is encapsulated using a molding compound. Note that the physical length of the mm-wave wirebonds is $1/\sqrt{\varepsilon_r}$ times smaller than the free space case (i.e., $\lambda_0/2$), where ε_r denotes the relative permittivity of the molding compound. The radiating elements of the antenna are not encapsulated by the molding compound. In contrast, the mm-wave wirebond interconnects and the via-based signal transition are encapsulated by the molding compound, hence it is important to select a molding compound with well-defined dielectric properties and a low loss tangent in the desired frequency range. Alternatively, if a hermetic lid is used, it will encapsulate the radiating elements of the antenna as well, which barring a special case, i.e., when the thickness of the lid is $\lambda_g/2$, would degrade the antenna radiation characteristics [BRG$^+$13]. In addition, a lid-based encapsulation entails a higher assembly effort than a molding encapsulation. The mm-wave signals are confined within the LTCC package and the DC and baseband signals are routed from the MMIC to a PCB (or vice versa), via the LFPB. The interconnects between the MMIC and LFPB are realized through standard wirebonds, without using any wirebond compensation technique. The interconnects between the LFPB and PCB are realized by soldering the LFPB's lead frame onto the surface of the PCB [10].
The manufacturing and assembly effort involved in implementing the horizontal configuration SiP is assessed on the basis of the following points.

1 Taking into consideration the typical LTCC manufacturing tolerances, it is proposed to manufacture the mm-wave module and LFPB separately. The mm-wave module requires only 4 layers of a high-frequency LTCC substrate, i.e., DuPont 9K7PV, for implementation. Therefore, the maximum possible layer misalignment is lower than the vertical-

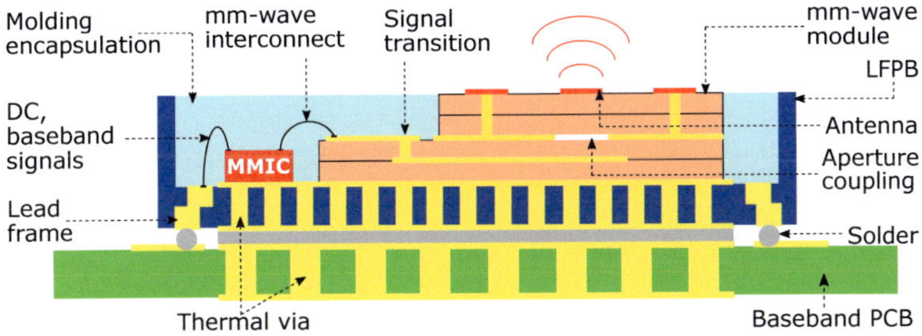

Figure 2.4: Horizontal-configuration SiP concept [10] © 2018 IEEE

configuration SiP concept. Additionally, state-of-the-art manufacturing techniques, e.g., laser ablation, are required to design the mm-wave module with a high resolution and accuracy. In contrast, for implementing the LFPB, it is sufficient to use a low frequency LTCC substrate, e.g., DuPont 951, since only DC and baseband frequency signals are routed through the LFPB. Consequently, the LFPB can be reliably manufactured by using the standard LTCC manufacturing process.

2 The heat generated by the active MMIC can be easily dissipated by the large metal pads (on the LFPB), which are connected by dozens of thermal vias. Unlike vertical-configuration SiP, no special structure, e.g., fluidic channel, is necessary for heat dissipation. The thermal management is easily achievable for MMICs with a power consumption of hundreds of milliwatts, a proof of which is shown in section 5.2.2.

3 The assembly effort of a horizontal-configuration SiP is broadly divided into the following steps: a) Placing and gluing the mm-wave module and the MMIC on the surface of the LFPB by using a thermally conductive adhesive b) Forming the $\lambda_g/2$ mm-wave wirebonds between the MMIC and the mm-wave module c) Forming the DC and baseband wirebonds between the MMIC and the LFPB's lead frame d) Filling the multistep cavity structure with a molding compound, followed by curing of the

molding compound e) Soldering the bottom side of the LFPB onto the surface of a baseband PCB.

2.3.3 An Overview of Contemporary mm-Wave SiPs

A brief summary of mm-wave SiPs realized in contemporary packaging technologies is given below. This is by no means an exhaustive list, but it gives an overview of the key packaging technologies, which are presently being used to realize mm-wave SiPs. A special focus is placed on the realization of mm-wave SiPs at or around 122 GHz in each case. Subsequently, a list of the recently demonstrated LTCC-based mm-wave SiPs is presented.

Thin film Antennas in Standard Plastic Package

In this concept, the off-chip antennas are realized using high-precision thin film technique on a single- or double-sided ceramic substrate (e.g., Al_2O_3) or an organic substrate (e.g., LCP). The antennas are integrated with an MMIC using wirebond or flip-chip interconnects and the complete assembly is placed in a standard plastic package, e.g., an open-cavity Quad Flat no-Lead (QFN) package. The plastic package acts as an input/output (I/O) interface between the MMIC and a baseband PCB. In addition, it facilitates thermal management and encapsulates the MMIC and antenna assembly. Different QFN based packaging concepts for mm-wave transceivers are shown in [ZB12]. One of these concepts is shown in Fig. 2.5. In [BRG$^+$13], a 122 GHz SiP, based on the same concept, is shown. In this SiP, a Tx antenna, an Rx antenna and two wirebond compensation networks are realized on two separate 100 µm thick Rogers Ultralam 3850 LCP substrates. The off-chip circuit includes narrow line width of 20 µm, which makes it necessary to use the thin film technology. The LCP substrate originally comes with a 19 µm Cu layer on top, which is etched away. Thereafter, the off-chip circuit is realized by using a 3 µm Cu layer coated with a 0.5 µm Au layer. The 0.5 µm Au layer is required to facilitate wirebonding between the thin film circuitry and the MMIC. The package is encapsulated using an Al_2O_3 lid with a thickness of $\lambda_g/2$ at 122 GHz. [BGY$^+$13] shows another 122 GHz SiP, in which two thin film antennas are integrated with a transceiver MMIC in a cavity-based package. In this package, the mm-wave interconnects between the antennas and MMIC are realized by means of flip-chip technology. The antennas face the die pad of the package, which in turn reflects the antenna beam to the top. Similarly to the previous SiP, an Al_2O_3 lid is used to encapsulate the package.

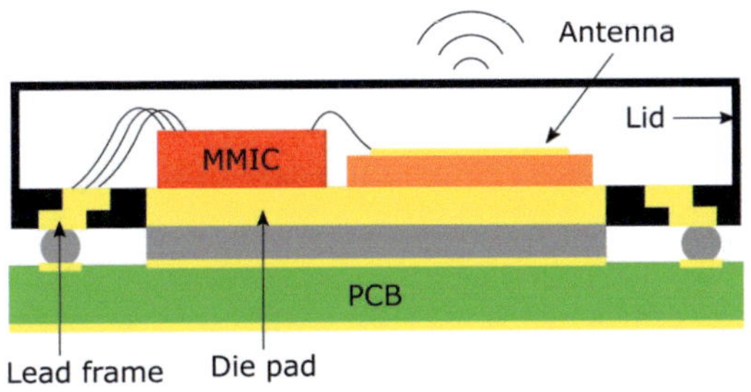

Figure 2.5: A QFN-based package concept [ZB12]

Embedded Wafer Level Ball Grid Array Package

The embedded wafer level ball grid array (eWLB) packaging technique is developed by Infineon Technologies, ST Microelectronics and STATS ChipPAC [HWP$^+$18]. A schematic of a standard eWLB package is shown in Fig. 2.6. In this concept, the I/O pins on an MMIC are fanned-out to a surrounding redistribution layer (RDL). The RDL consists of a dielectric layer and a Cu metallization layer, which are used to realize one or more antennas and other passive components (RDL with a maximum of two to three dielectric layers is possible). The MMIC and RDL are embedded in a molding compound with excellent dielectric characteristics in the desired mm-wave frequency range. The molded MMIC and RDL assembly is soldered face-down to a PCB by using a BGA. The PCB has a ground reflector on its top, which along with the RDL-based antenna constitutes a planar superstrate antenna configuration, thus achieving a wider bandwidth and a higher radiation efficiency. Up to now, several eWLB packages have been demonstrated at 77 GHz, such as [WWL$^+$12] and [FTH$^+$14]. Recently, an eWLB package at 122 GHz has been demonstrated in [FAS18]. In this package, a SiGe bipolar complementary metal-oxide semiconductor (BiCMOS) radar transceiver chip is integrated with a Tx and an Rx antenna array. The antenna arrays are designed on the RDL of the eWLB package and each antenna array consists of 43 rhombic elements. The measured gain of each antenna array is around 15.5 dBi and the eWLB package has a size of 18 mm×18 mm. The implementation of eWLB packages for frequencies beyond 100 GHz is influenced by the following two factors. First, the solder ball height should be strictly controlled, otherwise it alters the antenna-to-ground distance, thereby degrading the antenna radiation characteristics [HWP$^+$18]. Second, the antenna radiates through the molding compound, hence the dielectric characteristics of the molding compound should be well defined and its dielectric loss should be relatively low at frequencies beyond 100 GHz.

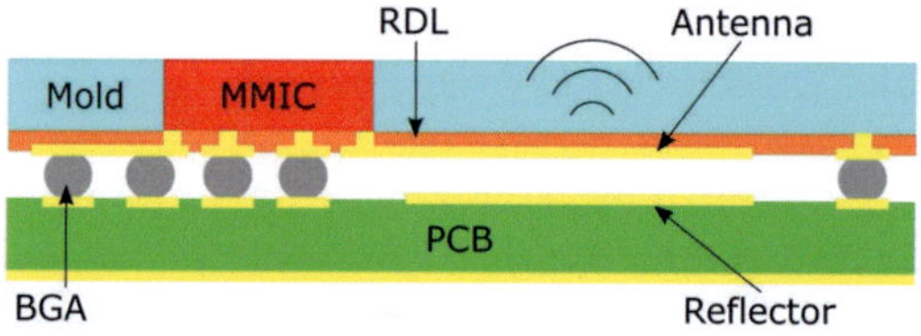

Figure 2.6: An eWLB package concept [WWL$^+$12]

Multilayer LCP Package

The electrical, mechanical and thermal properties of a typical high frequency LCP substrate have been already shown in Table 2.1. The main advantages of LCP over LTCC are as follows. First, a lower ε_r, which is useful for achieving a higher radiation efficiency. Second, the availability of thinner substrate layers of the order of 25 µm and third, a higher manufacturing resolution. On the downside, the LCP material is extremely soft and it has a lower processing temperature of around 300 °C, which causes difficulties while soldering an LCP-based SMP on a standard PCB. The integration of a 60 GHz transceiver MMIC in an LCP package is shown in [KBP+11] and [CAP11]. In [KBP+11], the LCP package consists of a Yagi-Uda antenna array, wirebonded to a GaAs Tx/Rx module. The Tx/Rx module is placed in a 200 µm deep cavity of the LCP package. It consists of a single-pole-double-throw switch, which enables switching between the Tx and Rx modes. In contrast, [CAP11] shows a multilayered LCP package with simultaneous Tx and Rx modes. In this case, the LCP package consists of two vertical dipole antennas and two dual-mode filters for improving the Tx-Rx isolation. The fully-integrated transceiver achieves a measured gain of 12 dBi and 22 dBi for the Tx and Rx mode, respectively. Recently, a 125.5 to 130.5 GHz complementary metal-oxide semiconductor (CMOS) multiple-input, multiple-output (MIMO) Tx chip integrated in an LCP package has been demonstrated in [LCP+16]. The LCP package consists of a 4×1 quasi-Yagi antenna array, which is connected to the CMOS Tx chip, using flip-chip interconnects. The LCP substrate has a thickness of nearly 50 µm and a minimum feature size of 20 µm. During the fabrication of this package, the LCP substrate was attached to a Si wafer in order to enhance the substrate rigidity and hence facilitate the manufacturing process.

Multilayer LTCC Packages To Date

In the past few years, a couple of LTCC-based mm-wave SiPs have been demonstrated, such as [SZC+08], [KLN+11], [BWMS13] and [ZGZ+15]. Most of these SiPs are designed for 60 GHz or 77 GHz frequency bands. Only [ZGZ+15] shows a 140 GHz LTCC SiP. It consists of a grid array antenna, an SL feed network, an SL-to-MS and an MS-to-GCPW signal transition. The off-chip circuitry is integrated with an InP power detector MMIC, using extremely short (100 µm) and flat wirebonds in ground-signal-ground

(GSG) configuration, which entail a high manufacturing complexity due to the following reasons. First, an extremely high-precision dicing is required for the LTCC package and the MMIC. Note that the dicing limit of LTCC modules is usually greater than 100 µm. Second, the placement of an MMIC within the LTCC package cavity should be done with a high accuracy. Third, producing flat wirebonds is extremely challenging. A significant deviation is observed between the simulated and measured antenna performance, which is attributed to the LTCC manufacturing tolerances. In addition, the thermal performance of the package as well as its encapsulation have not been discussed. The key performance parameters of LTCC-based mm-wave SiPs demonstrated in the recent years are shown in Table 2.4. The parameter η denotes the antenna efficiency.

Reference	Antenna type & feed	Bandwidth	Peak gain	η
[SZC$^+$08]	Yagi GSGSG	RC < -10 dB 60.6 - 62.9 GHz	6 dBi at 62 GHz	93%
[KLN$^+$11]	Phased array aperture-coupled	-	3 - 6 dBi at 60 GHz	-
[BWMS13]	Grid array LWG	RC < -12 dB 76 - 81 GHz	14 dBi at 79 GHz	46%
[ZGZ$^+$15]	Grid array SL-to-MS MS-to-GCPW	RC < -5 dB 136 - 148 GHz	17.6 dBi at 146 GHz	65%

Table 2.4: Recently demonstrated mm-wave LTCC SiPs

2.4 Summary

The electrical, thermal and mechanical properties as well as the metallization compatibility of an LTCC substrate are compared with two other multilayer technology substrates, namely, HTCC and LCP, thus establishing the suitability of LTCC for implementing a highly reliable and robust mm-wave SiP. After a thorough analysis of a wide range of commercial LTCC substrates, a high frequency LTCC substrate, namely, DuPont 9K7PV, which is specially designed for frequencies up to and beyond 100 GHz is selected for implementing the mm-wave components and systems in this work. Next, the manufacturing aspects of LTCC are investigated. First, an overview of the standard LTCC manufacturing process is given, which restricts the line/space width to 100 µm (typically), following which the state-of-the-art LTCC manufacturing techniques, capable of realizing narrow line/space width of the order of tens of micrometers are listed. Among these techniques, the laser ablation technique is used to realize narrow line/space width and microvias for the mm-wave components shown in this work. The smallest line width, space width and microvia diameter used in this work are 60 µm, 40 µm and 80 µm, respectively. Thereafter, the typical LTCC manufacturing errors, namely, shrinkage, warping, layer misalignment and cavity deformation are discussed along with their implications for a mm-wave design. Taking into consideration, the potential and pitfalls of LTCC manufacturing, two LTCC-based SiP concepts, namely, vertical- and horizontal-configuration SiP are presented. The concepts differ mainly in terms of the alignment of the MMIC and the mm-wave antenna. As a result, the requirements of the signal transition between the MMIC and antenna changes. An EM-coupled signal transition is proposed for implementing the vertical-configuration SiP and a via-based signal transition is proposed for implementing the horizontal-configuration SiP. Finally, a brief summary of the most significant mm-wave SiPs implemented in LTCC and other packaging technologies over the past few years is presented, thus establishing the significance and novelty of this research work.

3 mm-Wave SL-to-GCPW Signal Transitions in LTCC Technology

An SL-to-GCPW signal transition is a key component in the LTCC-based vertical- and horizontal-configuration SiP concepts, proposed in section 2.3.1 and 2.3.2, respectively. In an SL, the signal trace is buried between two semi-infinite ground planes, which leads to a negligible dispersion and an excellent shielding against the undesired EM interferences. Therefore, an SL is suitable for feeding an aperture-coupled antenna in a multilayered SiP. On the other hand, a GCPW is selected due to two reasons. First, the antenna should be connected to an MMIC, which in this case is a 122 GHz bistatic radar MMIC, whose mm-wave I/O pads are in GSG configuration. Second, the *S* parameter measurement of an mm-wave module is usually done by using GSG probes. An SL-to-GCPW signal transition can be realized in two ways, namely, a via-based signal transition and an EM-coupled signal transition. The LTCC-based SL-to-GCPW signal transitions demonstrated until now have been implemented up to a maximum frequency of 100 GHz. Moreover, all of them are via-based signal transitions. An EM-coupled SL-to-GCPW signal transition has not been implemented in LTCC technology. The focus of this chapter lies on the implementation of these two concepts in LTCC. It is naturally preferable to design a broadband signal transition, which operates from DC up to a frequency as high as possible. As far as the implementation of the LTCC-based horizontal- and vertical-configuration SiP is concerned, the MMIC selected in this work operates within the frequency range of 119.3 to 125.8 GHz. Therefore, the signal transition should have an RC less than -10 dB and a TC greater than -1 dB, at least within the frequency range of 115 to 130 GHz, after allowing some buffer on each side. The signal transitions presented in this chapter are in back-to-back configuration, i.e., GCPW-to-SL-to-GCPW, for the purpose of probe-based S parameter measurements.

3.1 Via-based SL-to-GCPW Signal Transition

Among the mm-wave SL-to-GCPW signal transitions demonstrated until now, [LGO05] showed a 55 to 70 GHz transition, using 7 layers of Ferro A6M substrate ($\varepsilon_r = 6.1$ and a layer thickness of 96.5 µm). The transition consists of a 50 Ω GCPW connected to an asymmetric SL by means of a stacked via (a stacked via is shown in Fig. 3.1a). The stacked via extends over 5 substrate layers. The signal via includes an additional catchpad, which compensates the inductive effect of the signal via, thereby achieving an RC less than −15 dB and a TC greater than −3 dB between 55 and 70 GHz. [Lee08] showed a 60 GHz transition using 6 layers of an LTCC substrate ($\varepsilon_r = 7$ and a layer thickness of 100 µm). In this transition, a 50 Ω GCPW is connected to a 50 Ω symmetric SL by means of three staggered vias (a staggered via is shown in Fig. 3.1a). In addition, two embedded air cavities are used to compensate the parasitic effects. The transition shows an RC less than −10 dB and a TC greater than −2 dB from 50 to 65 GHz. Most recently, [NMT$^+$13] showed a 10 MHz to 100 GHz transition using only 2 layers of DuPont 9K7PV substrate. In this transition, a 50 Ω GCPW and a 50 Ω symmetric SL are connected using the shortest possible signal via, i.e., the signal via extends over only one substrate layer. The transition shows an RC less than −10 dB and a TC greater than −2.5 dB from 10 MHz to 100 GHz. In this case, a broad bandwidth is achieved by means of two circular apertures of radius 127 µm, placed on the bottom ground plane. In addition, the signal via anti-pad is also tuned. The challenges involved in realizing a via-based signal transition with an extremely broad bandwidth extending from DC up to at least 130 GHz and even beyond if possible, are as follows.

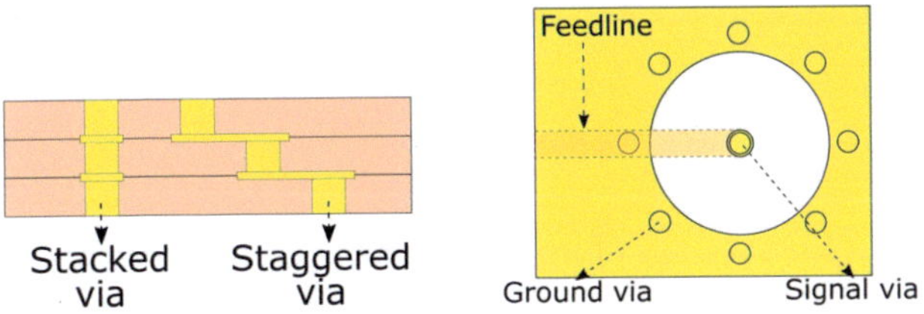

(a) Stacked and staggered vias

(b) A quasi-coaxial waveguide using vias

Figure 3.1: Types of via configuration

1 As the frequency increases, the signal path discontinuity caused by a via also increases, e.g., a stacked via extending over two layers of DuPont 9K7PV substrate, has an electrical length of around $\lambda_g/4.4$ at 122 GHz. Therefore, a signal via at such high frequencies becomes a significant source of reflection as well as radiation.

2 Replacing a large stacked via with multiple staggered vias is not an ideal solution, since staggered vias require additional via catchpads, which in turn lead to additional parasitic capacitances.

3 An alternative method used for designing a broadband signal transition is based on the principle of impedance matching. In this method, the signal via is surrounded by a cylindrical fence of ground vias, thus achieving a quasi-coaxial structure (see Fig. 3.1b) [ALHV10]. The signal via is fed by an MS, whose signal conductor is placed on the top metal layer (ML). The metal plane used for shorting the ground vias, also acts as the ground plane of the MS. If this concept is used to implement a 50 Ω quasi-coaxial structure in DuPont 9K7PV substrate and a signal via of 40 µm radius is used, then the center-to-center distance required between the signal via and each of the ground vias is 400 µm. In addition, the total substrate area occupied by the quasi-coaxial structure is

around 1 mm×1 mm, which is huge considering the fact that the entire SiP should not be more than a few millimeters in size.

A broadband via-based SL-to-GCPW signal transition operating at frequencies beyond 100 GHz is designed in this work, after taking the above mentioned points into consideration. The multilayered structure of the transition is shown in Fig. 3.2 [4]. The transition is made of three layers of DuPont 9K7PV substrate and four layers of Au metal. The ML 1 consists of 50 Ω GCPWs on either end. The bottom ground plane of GCPWs lies on ML 4. The signal conductor of each GCPW is connected to the signal conductor of an asymmetric SL (located on ML 2) by means of a signal via. The asymmetric SL is designed using a line width (w_{sl}) of 100 µm due to the manufacturing limitation of LTCC for buried substrate layers. Consequently, the asymmetric SL has an impedance of 41 Ω. The two ground planes located on ML 1 and ML 4 serve as upper and lower ground planes of the asymmetric SL. In addition, two rows of ground vias connecting the top and bottom ground planes are placed on either side of the GCPWs and the asymmetric SL. The ground vias help in suppressing an undesired parasitic parallel-plate mode. Moreover, the ground vias also serve as a return current path for the signal vias. Hence, the position of the ground vias influence the bandwidth of the signal transition. Lastly, two surface air cavities are placed directly below the signal via catchpads in order to reduce the parasitic capacitance effect. The air cavities extend over two substrate layers and the bottom ground plane on ML 4 [4].

3.1.1 Via Impedance and Lumped Element Model

A via can be characterized using a simple lumped element model, as shown in Fig. 3.3, if the rise time of a signal is nearly three times greater than the signal delay caused by the via [JG11]. The rise time of a signal is a function of its bandwidth. The 10 to 90% rise time of a signal (t_r) is calculated by using Eq. 3.1.

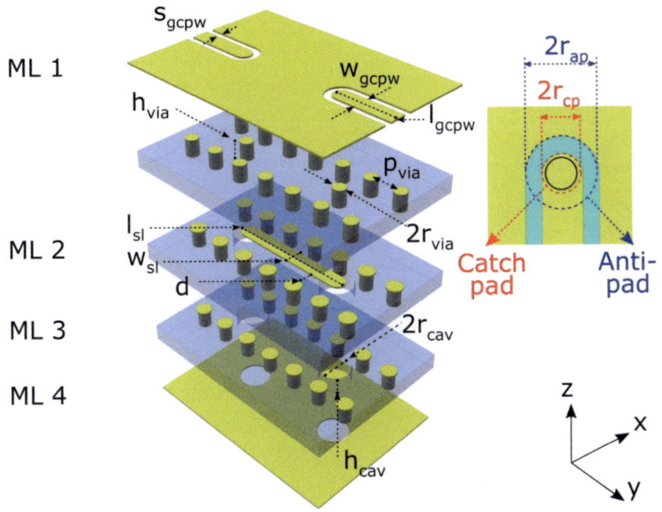

Figure 3.2: Multilayered structure of via-based SL-to-GCPW signal transition in back-to-back configuration [4]

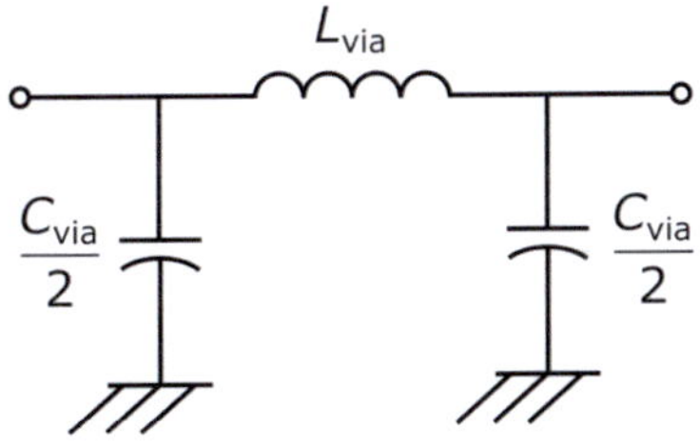

Figure 3.3: Lumped element model of a via

$$t_{\mathrm{r}}(\mathrm{ns}) \approx \frac{0.35}{B(\mathrm{GHz})} \tag{3.1}$$

Using a bandwidth (B) of 130 GHz (i.e., the desired upper frequency limit of the signal transition), t_{r} is calculated as 2.7 ps. The signal delay caused by a via (t_d) depends on the electrical length of the via and it is calculated by using Eq. 3.2 [JG11].

$$t_d \approx \frac{h_{via}\sqrt{\varepsilon_r}}{c_0} \tag{3.2}$$

The parameter h_{via} denotes the via height. Using $h_{via} = 105\,\mu m$ and $\varepsilon_r = 7$, t_d is calculated as 0.93 ps. Therefore, t_r is nearly three times greater than t_d, which implies that the via can be characterized using the lumped element model shown in Fig. 3.3. In this model, a via is represented by a series inductance L_{via} in the middle and two shunt capacitances $\frac{C_{via}}{2}$ on either side. The series inductance, L_{via}, of a via is calculated by using Eq. 3.3 [JG93].

$$L_{via}(\text{pH}) = \frac{\mu_0}{2\pi} h_{via} \left[\ln\left(\frac{2h_{via}}{r_{via}} \right) + 1 \right] \tag{3.3}$$

The parameters h_{via} and via radius, r_{via}, are in μm. The formula provides only a crude approximation of L_{via}. It assumes that the return signal current associated with the signal via flows through the ground vias, which are placed coaxially around the signal via. On substituting, $h_{via} = 105\,\mu m$ and $r_{via} = 40\,\mu m$, L_{via} is calculated as 55.8 pH. Next, the shunt capacitance, C_{via}, of a via is calculated by using Eq. 3.4 [JG93].

$$C_{via}(\text{fF}) = \frac{6.3\varepsilon_0\varepsilon_r h_{via} r_{via}}{r_{ap} - r_{via}} \tag{3.4}$$

The parameters h_{via}, r_{via} and via anti-pad radius, r_{ap}, are in μm. As in the previous case, the formula provides only a crude approximation of C_{via}. The equation is based on the assumption that $r_{via} = r_{cp}$ (where r_{cp} is the via catch-pad radius). In addition, the fringing field effect at the top and bottom of the via are ignored. On substituting $h_{via} = 105\,\mu m$, $r_{via} = 40\,\mu m$ and $r_{ap} = 91\,\mu m$ in Eq. 3.4, C_{via} is calculated as 32.3 fF. Finally, the characteristic impedance of a via, Z_{via}, is calculated by using Eq. 3.5 [JG11].

$$Z_{\text{via}} = \sqrt{\frac{L_{\text{via}}}{C_{\text{via}}}} \tag{3.5}$$

On substituting, L_{via}=55.8 pH and C_{via}=32.3 fF, Z_{via} is calculated as 41.6 Ω, which matches with the impedance of the asymmetric SL. Since, $Z_{\text{via}}<50\ \Omega$, the via results in a negative RC, which suggests that the via leads to an excess parasitic capacitance. This parasitic capacitance is mitigated by placing air cavities below the signal vias, as shown in Fig. 3.2. As previously mentioned, Eq. 3.3 and Eq. 3.4 provide only crude approximations of L_{via} and C_{via}, respectively. In reality, L_{via} and C_{via} are modified by the signal traces to which a signal via is attached as well as the ground vias placed in the vicinity of the signal via. Therefore, full-wave EM simulations of the via-based SL-to-GCPW signal transition are carried out in the Computer Simulation Technology (CST) Microwave Studio (MWS) software. Each of the following design parameters: r_{via}, h_{via}, r_{cp} and the distance between the ground vias and the asymmetric SL, d, is varied one at a time. The remaining design parameter values are shown in Table 3.1. The simulation result is analyzed in frequency domain in terms of S parameters (i.e., RC and TC). Thereafter, the S parameters are translated into a time domain reflectometry (TDR) signal plot by means of a post processing step in the CST MWS software. The TDR plots (shown in section 3.1.2) demonstrate time and impedance along x- and y-axis, respectively. The time corresponds to the total signal delay caused by the via-based SL-to-GCPW signal transition in back-to-back configuration (shown in Fig. 3.2). The TDR method is helpful in estimating the impedance discontinuities along the signal transition. The analysis is carried out for two simulation models, namely, an SL-to-GCPW signal transition with air cavity (see Fig. 3.2) denoted as 'AC' and an SL-to-GCPW signal transition without air cavity (same as Fig. 3.2 except cavities) denoted as 'NC'.

3.1.2 Parameter Study

Signal via radius

The parameter r_{via} is varied in three steps, namely, 40 µm, 50 µm and 60 µm (r_{cp}=70 µm in each case). The resulting TDR signal plot (see Fig. 3.4a) shows that as r_{via} decreases from 60 µm to 40 µm, Z_{via} increases for both AC and NC. A theoretical explanation for this phenomenon is as follows. As r_{via} decreases and h_{via} remains constant, L_{via} increases by a small amount, since r_{via} exists in Eq. 3.3 as an argument of natural logarithm. In contrast, C_{via} decreases by a large amount, since the numerator and denominator in Eq. 3.4 decrease and increase, respectively. Therefore, with decreasing r_{via}, the decrease in C_{via} is a dominant factor, which in turn leads to an increase in the Z_{via}. Moreover, AC provide higher values of Z_{via} than NC, which indicates a reduction in the parasitic capacitance due to the presence of air cavities. Further, r_{via}=40 µm (for AC model) shows the least amount of impedance discontinuity in time domain. Consequently, in frequency domain, it shows the lowest RC (see Fig. 3.4b) and the highest TC (see Fig. 3.4c), i.e., the RC is less than −10 dB and the TC is greater than −1.2 dB up to a frequency of 134.8 GHz.

Substrate height and signal via height

Since the signal via passes through only one substrate layer (see Fig. 3.2), h_{sub}=h_{via}. As mentioned in section 2.2.3, the LTCC shrinkage tolerance leads to a variation in the substrate height. Therefore, both h_{sub} and h_{via} are simultaneously varied in the following three steps: 105 µm, 115 µm and 125 µm. From the TDR signal plot shown in Fig. 3.5a, it is observed that the impedance of the asymmetric SL increases with an increase in h_{sub} and Z_{via} increases with an increase in h_{via}. The latter can be understood in the following manner. Both L_{via} and C_{via} increase with an increase in h_{via}, but the increase in L_{via} dominates over the increase in C_{via} (see Eq. 3.3 and Eq. 3.4). Hence, Z_{via} increases with an increase in h_{via}. Further, the impedance curves of the AC model show higher values of Z_{via} than the corresponding impedance curves of

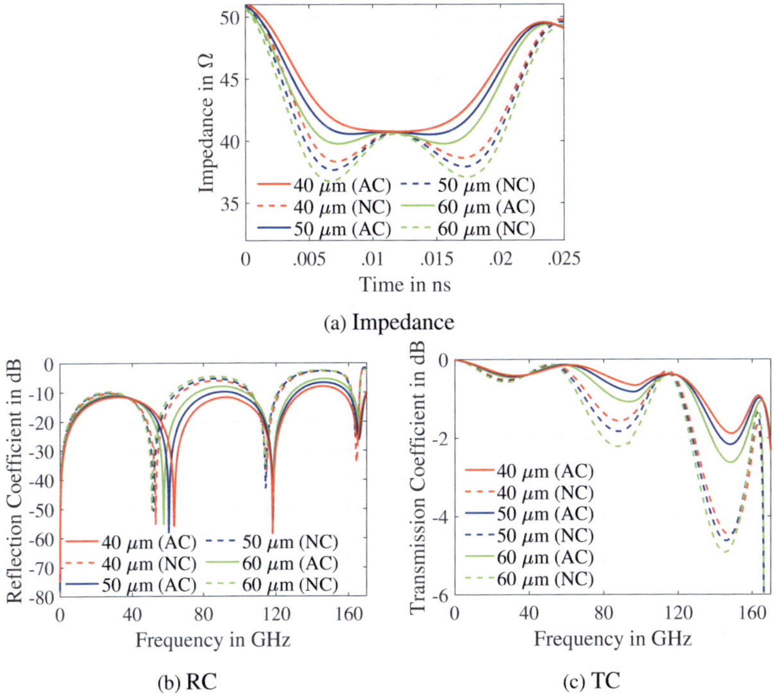

Figure 3.4: Simulated influence of varying signal via radius on the performance of via-based SL-to-GCPW signal transition

the NC model. This verifies that the AC model helps in reducing the parasitic capacitance. On observing the impedance curves of the AC model, it is seen that the value of Z_{via} is well matched to the impedance of the asymmetric SL. In frequency domain, $h_{\text{sub}}=h_{\text{via}}=105$ µm shows the best performance in terms of RC and TC. Fig. 3.5b and Fig. 3.5c show that the RC is less than -10 dB and the TC is greater than -2 dB up to a frequency of 170 GHz.

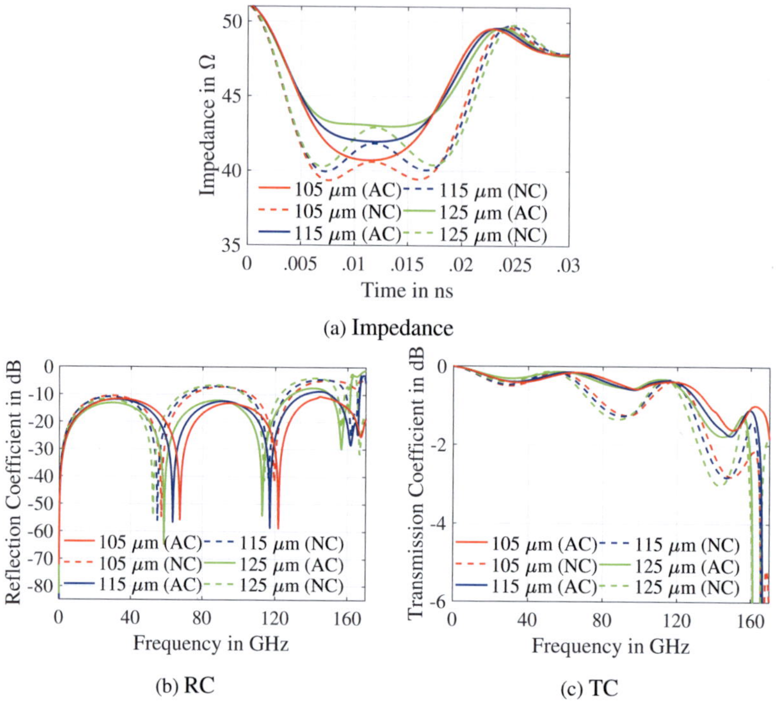

Figure 3.5: Simulated influence of varying signal via height on the performance of via-based SL-to-GCPW signal transition

Signal via catchpad radius

The parameter r_{cp} is varied in the following three steps: 50 µm, 60 µm and 70 µm (r_{via}=40 µm in each case). From the resulting TDR signal plot shown in Fig. 3.6a, it is observed that Z_{via} decreases with an increase in r_{cp}. This phenomenon can be understood as follows. As seen in Eq. 3.3, L_{via} is independent of r_{cp}. On the other hand, Eq. 3.4 shows that C_{via} increases with an increase in r_{cp}. Note that the equation assumes $r_{via}=r_{cp}$, therefore the numerator and denominator increase and decrease, respectively. Consequently, Z_{via} decreases with an increase in r_{cp}. Further, on comparing the impedance curves of the AC and NC models, it is seen that the AC model shows only a minor change in Z_{via} with varying r_{cp}. The same is visible in the frequency domain results as well. Fig. 3.6b and Fig. 3.6c show that the AC model with r_{cp}=50 µm achieves an RC less than −10 dB and a TC greater than −2 dB up to a frequency of 170 GHz.

Ground via position

The influence of the ground via position is investigated by varying the edge-to-edge distance between the asymmetric SL and one of the ground via rows (see parameter d in Fig. 3.2) in the following three steps: 100 µm, 120 µm and 140 µm. In this case, 100 µm is the minimum distance that can be manufactured using LTCC. The TDR signal plot (see Fig. 3.7a) shows that the impedance of the asymmetric SL and Z_{via} increase with an increase in the value of d. Note that the L_{via} and C_{via} equations shown in Eq. 3.3 and Eq. 3.4, respectively, assume a coaxial configuration of ground vias around the signal via. Therefore, in this case the impedance change cannot be explained on the basis of the via lumped element model. The frequency domain results (see Fig. 3.7b and Fig. 3.7c) show that up to 134 GHz, the RC and TC curves for the AC model remain constant with varying d. Thereafter, the curves corresponding to a distance of 140 µm and 120 µm show a sharp resonance at 144 GHz and 158 GHz, respectively. A similar pattern is observed in the RC and TC curves of the NC model as well. Moreover, for the NC model, the sharp resonances occur at frequencies lower than in the AC model. For the AC model, d=100 µm achieves an RC less than −10 dB and a TC greater than −2 dB up to a frequency of 170 GHz. Based on the theoretical and parametric study analysis, the design parameters of the via-based signal transition (shown in Fig. 3.2) are optimized. The optimized values of the design parameters are given in Table 3.1.

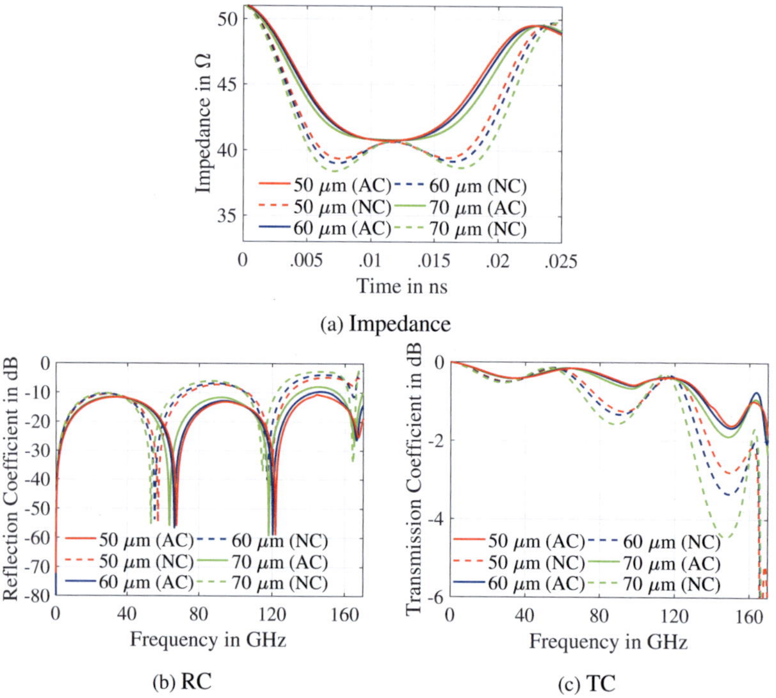

Figure 3.6: Simulated influence of varying signal via catchpad radius on the performance of via-based SL-to-GCPW signal transition

3.1.3 Prototype and Measurement Result

Prototypes of both AC and NC variants are manufactured. The top and bottom MLs of the prototypes are inspected by using a high-resolution microscope. The manufactured dimensions of the following parameters: w_{gcpw}, s_{gcpw} and r_{cav} are compared with their values in the initial simulation model, shown in Table 3.1. The microscope images of the top and bottom MLs of an AC prototype are shown in Fig. 3.8 [4]. Subsequently, the prototypes are analyzed using a synchrotron-based X-Ray microtomography facility [FaKIoT] with a high spatial resolution of 2.44 μm. The X-ray images are analyzed using

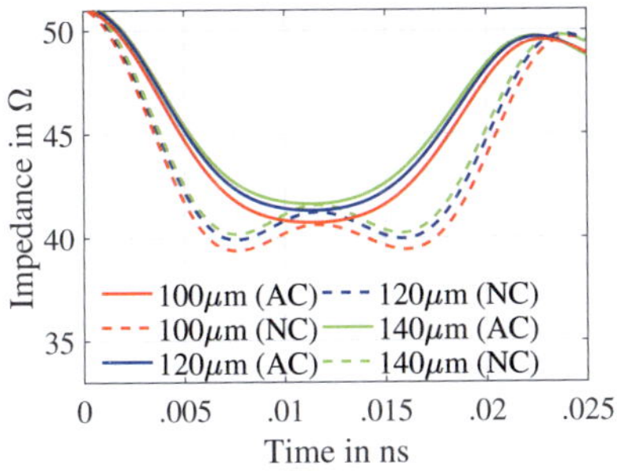

(a) Impedance

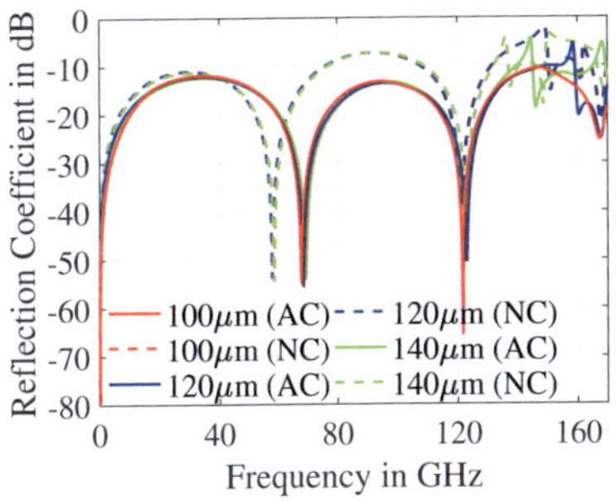

(b) RC

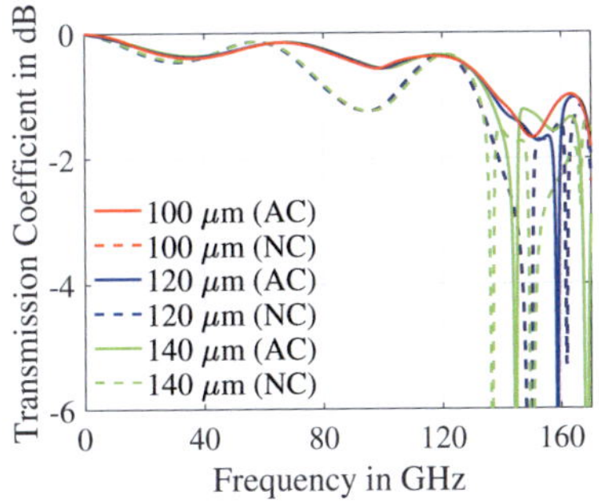

(c) TC

Figure 3.7: Simulated influence of varying ground via position on the performance of via-based SL-to-GCPW signal transition

an open-source image processing software named Fiji [SCF+12]. The X-ray images of two cross-sectional planes are shown in Fig. 3.9. Fig. 3.9a shows the X-ray image of a plane passing through the middle of a signal via and an air cavity. It is observed that the air cavity, which is designed as a cylinder in the initial simulation model, is manufactured as a truncated cone with variable diameter along its height. Fig. 3.9b shows the X-ray image of a plane passing through the middle of the asymmetric SL. In this image, the signal conductor of the asymmetric SL and the ground via catchpads on either side are visible. The Fiji software is used to measure the following dimensions from these X-

Ray images: r_{via}, h_{via}, r_{cav}, h_{cav}, and w_{sl}. The dimensions are measured for four AC prototypes, based on which the mean and standard deviation of the manufactured dimensions are calculated for each parameter. These values are shown in Table 3.2.

Parameter (see Fig. 3.2)	Value (µm)	Parameter (see Fig. 3.2)	Value (µm)
w_{gcpw}	100	r_{ap}	91
s_{gcpw}	41	w_{sl}	100
l_{gcpw}	400	l_{sl}	800
r_{via}	40	d	100
h_{via}	105	r_{cav}	100
p_{via}	200	h_{cav}	210
r_{cp}	50		

Table 3.1: Design parameter values of via-based SL-to-GCPW signal transition [4]

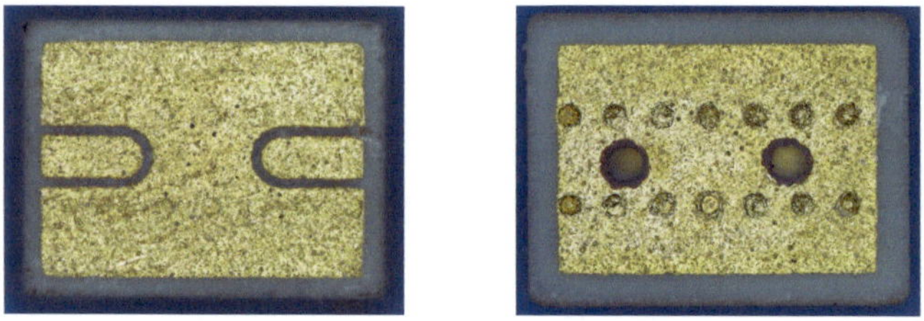

(a) Top layer (b) Bottom layer

Figure 3.8: Microscope images of via-based SL-to-GCPW signal transition with air cavities [4]

The standard deviation values of the dimensions are less than 8 µm, which indicates a good reproducibility of the LTCC process. Subsequently, the initial simulation models of both AC and NC models are modified by using the mean values of the manufactured dimensions, shown in Table 3.2. The remaining parameter values are kept the same as in the initial simulation model.

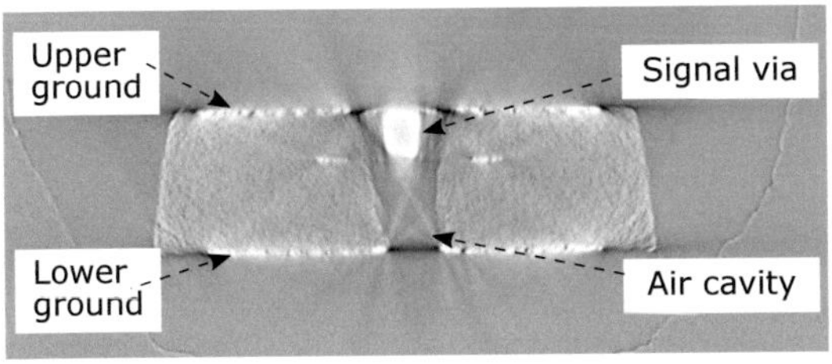

(a) Cross-section of one of the signal vias and air cavities

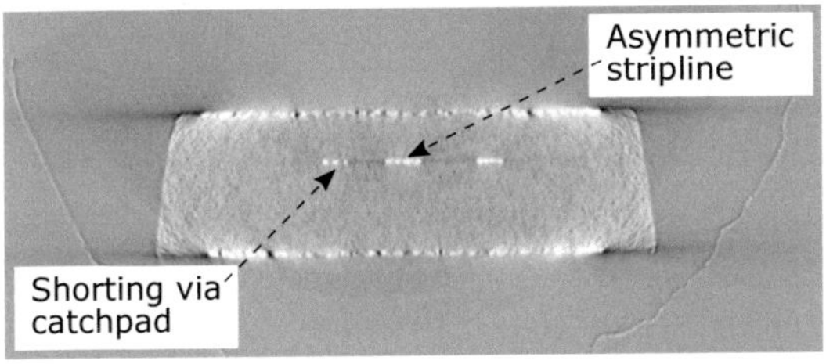

(b) Cross-section of the asymmetric SL

Figure 3.9: X-ray images of via-based SL-to-GCPW signal transition with air cavities

The S parameters of an AC and an NC prototype are measured using a Cascade MicroTech probe station. The measurement is done in two frequency ranges, namely, 10 MHz to 110 GHz and 110 to 170 GHz, using GSG probes with a pitch of 150 µm. The measurements are preceded by a short-open-load-through (SOLT) calibration in 10 MHz to 110 GHz frequency range and a thru-reflect-line (TRL) calibration in 110 to 170 GHz frequency range. The measurement and simulation results of the AC and NC prototypes are discussed below. For each prototype, the measured RCs at both ports (i.e., the measured S_{11} and the measured S_{22} denoted as Meas. S_{11} and Meas. S_{22}, respectively) are compared with the initial simulation RC [denoted as Sim.(init) S_{11}] as well as the modified simulation RC [denoted as Sim.(mod) S_{11}]. Note

Parameter (see Fig. 3.2)	Simulation (μm)	Manufactured Mean (μm)	Manufactured Std. Dev. (μm)
h_{cav}	210	217.47	2.08
r_{cav} (max)	100	95.46	7.88
r_{cav} (min)	-	65.57	2.34
r_{via}	40	44.38	3.78
h_{via}	105	121.70	4.15
w_{sl}	100	87.84	5.98
w_{gcpw}	100	178.25	2.06
s_{gcpw}	41	39.75	2.22

Table 3.2: X-ray based measurement analysis of via-based SL-to-GCPW signal transition with air cavities

that the initial and modified simulation models are perfectly symmetrical due to a back-to-back configuration, therefore the simulated S_{22} is the same as the simulated S_{11}. In terms of TC, the measured TC at both ports (i.e., the measured S_{21} and the measured S_{12} denoted as Meas. S_{21} and Meas. S_{12}, respectively) are compared with the initial simulation TC [denoted as Sim.(init) S_{21}] as well as the modified simulation TC [denoted as Sim.(mod) S_{21}]. Due to the reciprocity of a passive component, the S_{21} and S_{12} curves are identical.

Via-based signal transition with air cavity

In terms of the RC (see Fig. 3.10a [4]), the first two resonances observed in the Meas. S_{11} and Meas. S_{22} curves occur at around 38 GHz and 85 GHz, which are visible in the Sim.(mod) S_{11} curve as well. The third resonance is seen at around 130 GHz in the measurement, which coincides with the resonance observed in the Sim.(init) S_{11} curve. In addition, the Meas. S_{11} and Meas. S_{22} are less than -10 dB up to a frequency of around 150 GHz. In terms of the

TC (see Fig. 3.10b [4]), the Meas. S_{21} curve shows a good agreement with the Sim.(mod) S_{21} curve. The Meas. S_{21} sharply degrades at around 160 GHz, which is clearly visible in the Sim.(mod) S_{21} as well. This effect is not seen in the Sim.(init) S_{21}. Further, the Meas. S_{21} is greater than -2 dB up to a frequency of 153 GHz. Consequently, for the single-ended signal transition, the measured TC is greater than -1 dB up to a frequency of 153 GHz. Overall, the measurement results show a better agreement with the modified simulation results, thus validating the X-ray analysis.

Via-based signal transition without air cavity

In terms of the RC (see Fig. 3.10c), the Meas. S_{11} and Meas. S_{22} curves show resonances at 33 GHz and 80 GHz, which lie closer to the resonances seen in the Sim.(mod) S_{11} curve. The Meas. S_{11} and Meas. S_{22} are less than -8.5 dB up to a frequency of 145 GHz. Therefore, in comparison to the via-based signal transition with air cavity (AC), the impedance matching shows a degradation of 1.5 dB and the frequency range is lower by nearly 5 GHz. In terms of the TC (see Fig. 3.10d), the Meas. S_{21} shows a good agreement with the Sim.(mod) S_{21}. The Meas. S_{21} sharply degrades at around 154 GHz, which is visible in the Sim.(mod) S_{21} curve as well. The Meas. S_{21} is greater than -2 dB up to a frequency of 147 GHz. Therefore, the bandwidth in this case is 6 GHz lower than the via-based signal transition with air cavity (AC). A comparison of the measured S parameters of the via-based signal transitions with air cavity (AC) and without air cavity (NC) at certain frequencies are shown in Table 3.3. It is observed that the performance of the via-based signal transition with air cavity (AC) is better than the via-based signal transitions without air cavity (NC) over the entire frequency range of DC to 150 GHz.

3.2 EM-Coupled SL-to-GCPW Signal Transition

In an SL-to-GCPW signal transition, if the signal conductors of a GCPW and an SL are separated by more than one layer of DuPont 9K7PV substrate

($\varepsilon_r = 7$ and $h_{\text{sub}} = 105\,\mu\text{m}$), then the via-based approach cannot be used to realize a signal transition at frequencies above 100 GHz. This is mainly due to the following manufacturing limitations. First, the minimum via radius is 40 µm. Second, the via catchpad radius should be at least 10 µm greater than the via radius itself. Third, it is difficult to manufacture embedded air cavities, since they are susceptible to deformation and sagging during the co-firing process. Fourth, the minimum via-to-line spacing is 100 µm. Therefore, an alternative approach based on EM coupling is used to realize an SL-to-GCPW signal transition.

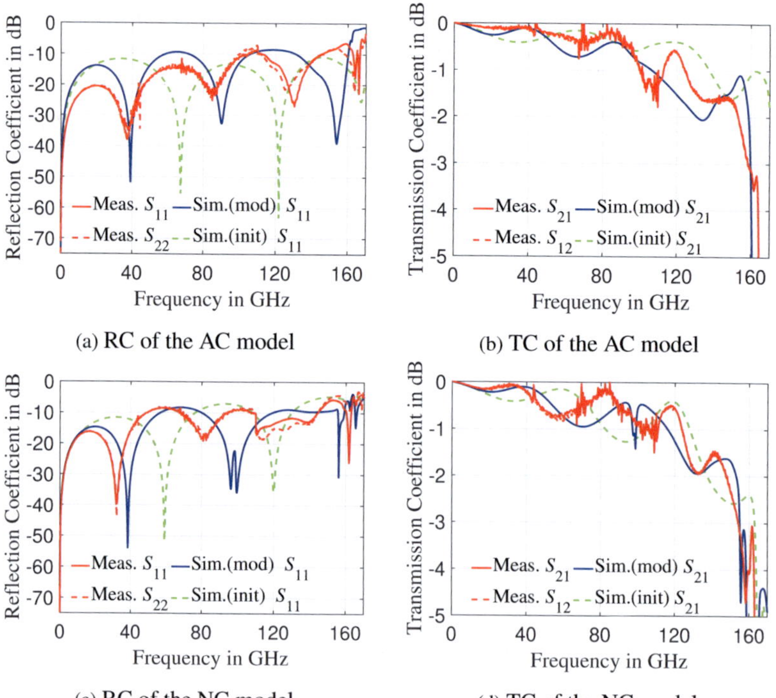

(a) RC of the AC model

(b) TC of the AC model

(c) RC of the NC model

(d) TC of the NC model

Figure 3.10: Measured and simulated S parameters of via-based SL-to-GCPW signal transition with air cavity (AC) [4] and without air cavity (NC)

Frequency (GHz)	Measured RC (dB)		Measured TC (dB)	
	with air cavity (AC)	without air cavity (NC)	with air cavity (AC)	without air cavity (NC)
25	-21.40	-19.07	-0.075	-0.090
50	-17.81	-9.82	-0.24	-0.70
75	-15.84	-13.50	-0.40	-0.50
100	-11.05	-9.02	-1.02	-0.92
125	-17.07	-12.25	-1.04	-1.20
150	-8.80	-6.43	-1.70	-2.40

Table 3.3: Performance comparison of back-to-back via-based SL-to-GCPW signal transition with air cavity (AC) [4] and without air cavity (NC)

Unlike the via-based signal transition, the EM-coupled signal transition presented in this section, consists of a GCPW and an SL, whose signal conductors are separated by two layers of DuPont 9K7PV substrate and an intermediate ground plane. Such a configuration is essential for realizing the vertical-configuration SiP proposed in section 2.3.1. The EM coupling between two vertically-stacked TLs can be achieved by means of either proximity- or aperture-based EM coupling. In a proximity-based EM coupling, the signal conductors of the two TLs are coupled through a dielectric substrate, whereas in an aperture-based EM coupling, the signal conductors are coupled through an aperture, placed in an intermediate common ground plane. Among previous works, [JM87] showed an EM-coupled signal transition between two vertically-stacked coplanar waveguides (CPWs). In the coupling region of this transition, the slot width of both launching and receiving CPW is increased and the CPWs are terminated in a conventional open circuit (OC). The transversal size of the CPWs required for implementing the signal transition is dependent on the vertical separation of the two CPWs. The longitudinal size of the CPWs in the coupling region is $\lambda_g/4$ at the operating frequency. Further, EM-coupled CPW-to-MS and MS-to-CPW signal transitions are demonstrated in [BJ89] and [ZM04], respectively. In [ZM04], the coupling region of the transition consists of an OC CPW superimposed over an OC MS. The CPW twin slots, the CPW signal conductor and the MS signal conductor are widened in order to achieve a tight series capacitive coupling. A similar approach is shown in [BJ89] as well. Both signal transitions have been implemented at frequencies below 10 GHz. A mm-wave EM-coupled signal transition in LTCC has not been demonstrated to date.

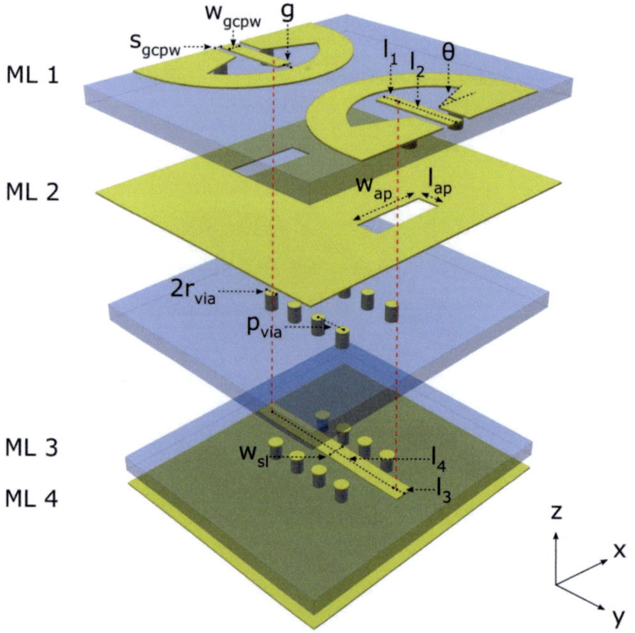

Figure 3.11: Multilayered structure of EM-coupled SL-to-GCPW signal transition in back-to-back configuration [21] © 2018 IEEE

The EM-coupled SL-to-GCPW signal transition investigated in this work, is shown in Fig. 3.11 [21]. The signal transition consists of three DuPont 9K7PV substrate layers and four Au MLs. The ML 1 consists of two 50 Ω GCPWs on either end. For each GCPW, two ground vias are used to connect the lateral ground planes to the bottom ground plane on ML 2, thereby suppressing an undesired coupled slotline mode. The twin slots of each GCPW are gradually widened. In addition, a modified OC is formed by terminating the GCPW signal conductor a short distance before a semiannular ground ring. The ML 2 consists of an intermediate ground plane, which is common to the GCPWs and a symmetric SL. The intermediate ground plane consists of two rectangular apertures, which enable EM coupling between the GCPWs and SL. The ML 3 includes the SL signal conductor and two rows of ground vias placed on its either side. The ground vias connect the top and bottom ground planes of the SL (placed on ML 2 and ML 4, respectively), thereby suppressing an undesired parallel plate mode [21].

3.2.1 Coupling Mechanism

A 50 Ω GCPW is realized by using a substrate thickness of $h_{\text{sub}} = 105\ \mu\text{m}$, a signal conductor width of $w_{\text{gcpw}} = 100\ \mu\text{m}$ and a slot width of $s_{\text{gcpw}} = 50\ \mu\text{m}$. Since the s_{gcpw} is relatively narrow, the dominant mode of propagation is an (even) CPW mode, in which the electric field (E-field) lines are concentrated in the twin slots of the GCPW. This is shown by the E-field distribution in Fig. 3.12a. As the GCPW twin slots are gradually widened (i.e., the parameter s_{gcpw} is increased and the parameter θ shown in Fig. 3.11 is also varied), the CPW mode changes into an MS mode. As a result, the E-field lines are no longer concentrated in the GCPW twin slots, instead they are directed from the GCPW signal conductor towards the GCPW bottom ground plane (or vice versa) [Sim01] [Wol06]. Hence, EM coupling is achieved between the two TLs. Fig. 3.12b shows the E-field distribution in the XZ plane (see xyz coordinate system in Fig. 3.11) of the coupling region, i.e., the XZ plane passes through the middle of one of the ground plane apertures. The same principle has been verified by means of a parameter study. The design parameter

θ (see Fig. 3.11) is varied in the following three steps: 90°, 52.7° and 20.6° (see Fig. 3.13c). The corresponding RCs and TCs of the EM-coupled signal transition are shown in Fig. 3.13a and Fig. 3.13b, respectively. The remaining design parameter values are optimized on the basis of full-wave EM simulations carried out in the CST MWS transient solver. These values are shown in Table 3.4.

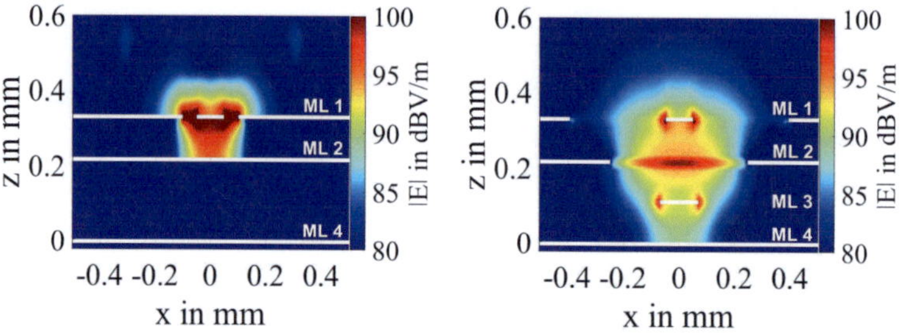

(a) 50 Ω GCPW with narrow twin slots

(b) Coupling region with wide twin slots and ground plane aperture

Figure 3.12: Simulated E-field of EM-coupled SL-to-GCPW signal transition

For θ=90°, the GCPW structure on ML 1 resembles a conventional CPW OC. An ideal OC has an RC of 1, i.e., the incident wave is totally reflected. In contrast, in an actual CPW OC, a part of the incident wave is transmitted beyond the signal conductor discontinuity. As a result, fringing fields exist in the gap (g) between the GCPW signal conductor and the semiannular ground ring. Moreover, a certain amount of radiation loss is also bound to occur. At 122 GHz, the RC and TC for θ=90° are −1.2 dB and −9 dB, respectively. As the GCPW twin slots are widened by decreasing θ, both RC and TC show an improvement. For θ=52.7°, the RC and TC are −11.16 dB and −1.63 dB, respectively. Finally, for θ=20.6°, the RC and TC are −22 dB and −0.99 dB, respectively, at 122 GHz. The RC and TC curves corresponding to θ=20.6° show the optimized performance of the EM-coupled SL-to-GCPW signal transition. The simulation result shows that the RC is less than −14 dB and the TC is greater than −2 dB in the frequency range of 114.5 to 132 GHz, i.e., 14.2% relative bandwidth. Therefore, for a single-ended EM-coupled SL-to-GCPW signal transition, the TC is greater than −1 dB in this frequency range. The radiation loss of the signal transition is obtained by using farfield boundary conditions in the CST simulation. The resultant plot is shown in Fig. 3.13d. The radiation loss between 120 and 130 GHz is close to 10%. The radiation loss value is reasonable considering the high frequencies involved.

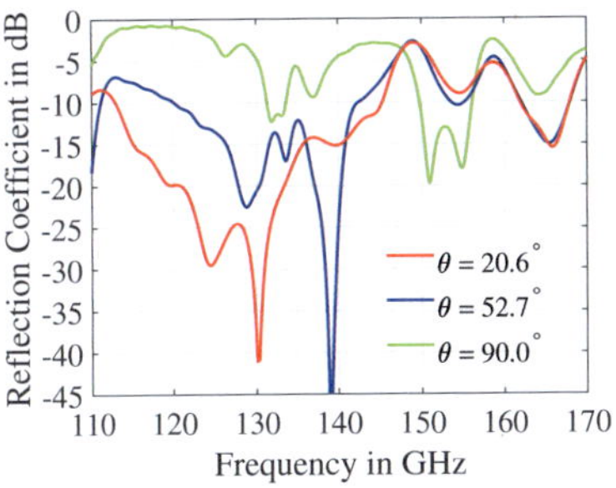

(a) Simulated RC showing the influence of modifying the GCPW open circuit

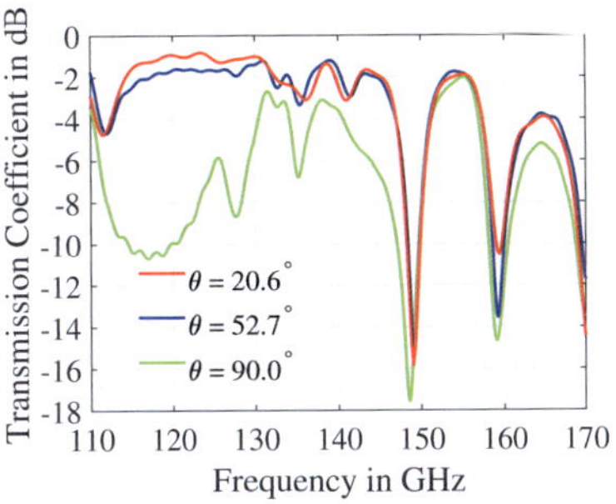

(b) Simulated TC showing the influence of modifying the GCPW open circuit

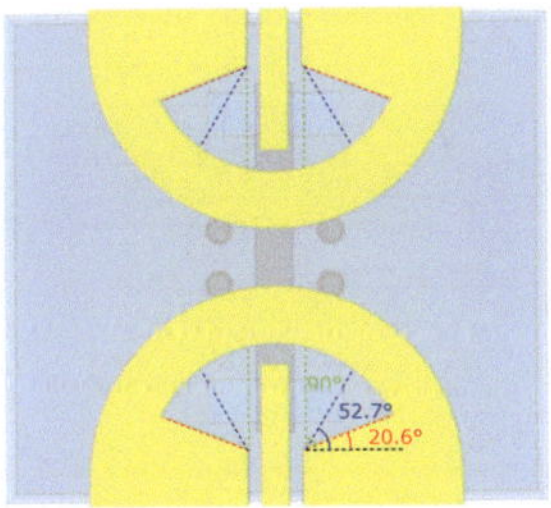

(c) Opening angle of the GCPW twin slots (θ) varied in steps

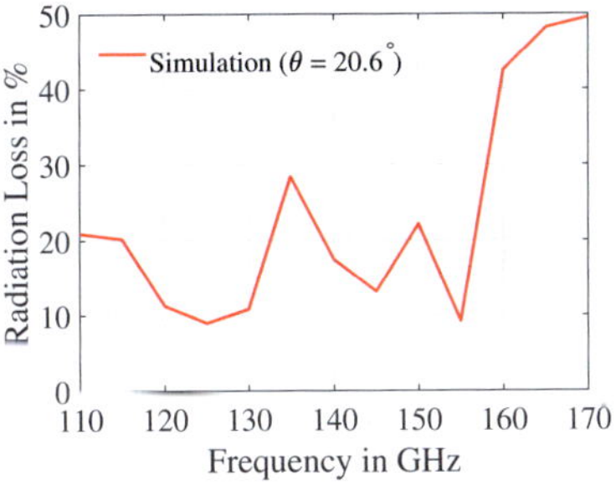

(d) Simulated radiation loss for the optimized GCPW open circuit

Figure 3.13: Optimization of EM-coupled SL-to-GCPW signal transition

The modified OC used in the signal transition resembles an OC MS stub, for which the radiation loss can be calculated by using Eq. 3.6 [Lew60].

$$P_{\text{rad}} = 60\left(\frac{2\pi h_{\text{sub}}}{\lambda_0}\right)F(\varepsilon_{\text{eff}}) \tag{3.6}$$

The parameters P_{rad}, F and ε_{eff} denote the radiated power for a unit incident current wave, the radiation factor and the effective dielectric constant, respectively. The value of $F(\varepsilon_{\text{eff}})$ depends on the type of discontinuity. $F(\varepsilon_{\text{eff}})$ for an OC is nearly three times larger than for a matched termination [Lew60]. The equation shows that the radiation loss increases with the frequency. An EM coupled MS-to-CPW signal transition operating from 3.2 to 11.2 GHz shows a loss of up to 4% [ZM04] and therefore the loss percentage observed in this case seems to be justified. In addition, the radiation loss peaks seen at higher frequencies, e.g., at 135 GHz and 150 GHz depend on the dimensions of the ground plane aperture located on ML 2.

Parameter (see Fig. 3.11)	Value (µm)	Parameter (see Fig. 3.11)	Value (µm)
w_{gcpw}	100	l_{ap}	152
s_{gcpw}	50	r_{via}	40
g	76	p_{via}	200
l_1	136	l_3	122
l_2	360	l_4	136
w_{ap}	494	w_{sl}	140

Table 3.4: Design parameter values of EM-coupled SL-to-GCPW signal transition [21] © 2018 IEEE

3.2.2 Prototype and Measurement Result

Multiple prototypes of the EM-coupled signal transition are manufactured. Fig. 3.14a shows the microscope image of the top ML of one of the prototypes [21]. The top ML is inspected by using a high-resolution microscope and the manufactured dimensions of the following design parameters: w_{gcpw}, s_{gcpw} and g are measured for multiple prototypes. The mean values of the measured s_{gcpw} and g show a deviation of -36% and $+29\%$, respectively, in comparison with their simulated values. The manufactured GCPW signal conductor shows a step discontinuity. The average measured value of w_{gcpw} varies between 124 µm (on the 50 Ω GCPW side) and 94 µm (on the modified OC side). Therefore, the maximum and minimum values of the manufactured w_{gcpw} differ from the initially simulated value (100 µm) by $+24\%$ and -6%, respectively. Subsequently, the buried substrate and MLs are analyzed by using X-ray based microtomography (with a resolution of 2.44 µm along x, y and z directions). A top view X-ray image of a signal transition prototype (see Fig. 3.14b [21]) shows that the SL signal conductor is misaligned in both transversal and longitudinal directions. The exact amount of the misalignment is determined by observing the X-ray images of the signal transition along the following two (orthogonal) cross sections. First, Fig. 3.14c [21] shows the X-ray image of the XZ plane passing through the midpoint of one of the rectangular apertures. In this image, two dimensions are measured, namely, δx_1 and δx_2. δx_1 denotes the distance between the midpoint of the GCPW signal conductor and the rectangular aperture edge and δx_2 denotes the distance between the midpoint of the SL signal conductor and the rectangular aperture edge. δx_1 and δx_2 are measured to be around 260 µm and 220 µm, respectively. Consequently, the difference of δx_2 and δx_1, which indicates the XZ-plane misalignment between the GCPW and SL signal conductors is calculated as 40 µm. Second, Fig. 3.14d [21] shows the X-ray image of the YZ-plane passing through the midpoints of both rectangular apertures. In this image, two dimensions are measured, namely, δy_1 and δy_2. δy_1 denotes the distance between the OC edge of the GCPW signal conductor and the rectangular aperture edge and δy_2 denotes the distance between the SL signal conductor edge and the rectangular aperture edge. Since, the prototype is asymmetric (due to LTCC manufacturing defect), δy_1 and δy_2 are measured on both left and right sides.

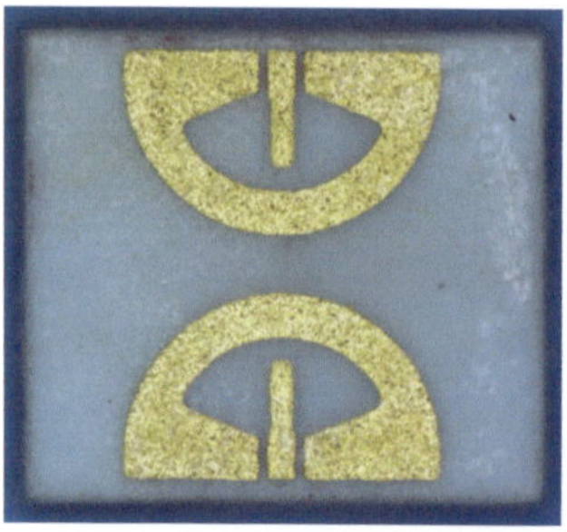

(a) Microscope image (Top view)

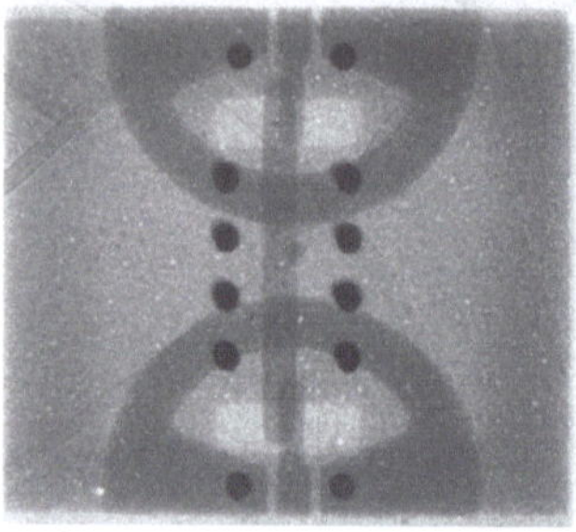

(b) X-ray image (Top view)

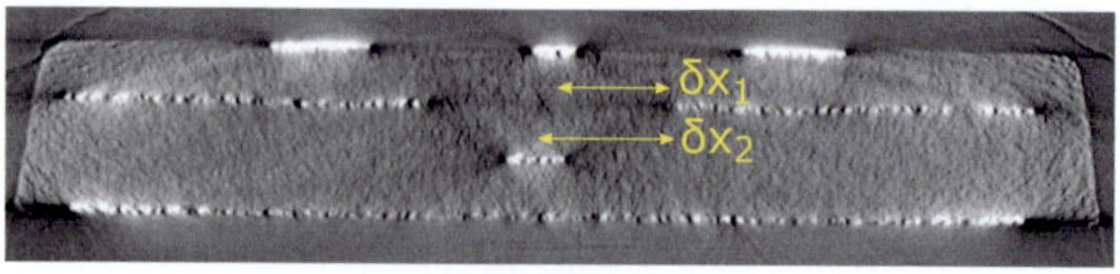

(c) X-ray image (XZ plane cross section)

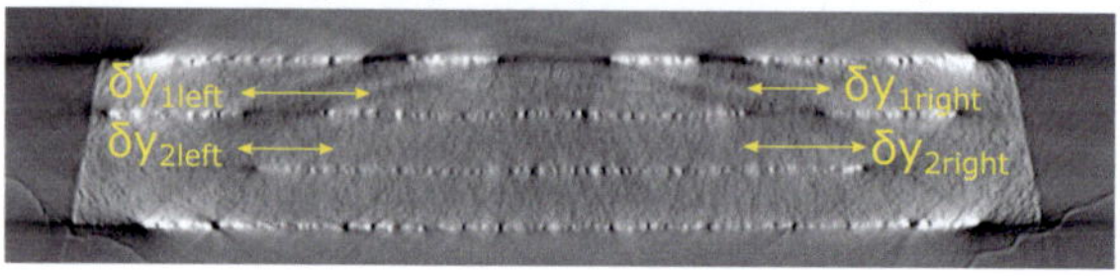

(d) X-ray image (YZ plane cross section)

Figure 3.14: A prototype of EM-coupled SL-to-GCPW signal transition [21] © 2018 IEEE

The measured values of $\delta y_{1\text{left}}$, $\delta y_{1\text{right}}$, $\delta y_{2\text{left}}$ and $\delta y_{2\text{right}}$ along with the initial simulation values are shown in Table 3.5 [21]. After comparing the manufactured dimensions and the initial simulation values in the YZ-plane, it is observed that the GCPW and SL signal conductors are longitudinally misaligned by approximately 50 µm. Besides the misalignment error analysis, the following buried layer dimensions are also measured using X-ray images: rectangular aperture length (l_a), rectangular aperture width (w_a), SL stub length (l_3) and SL signal conductor width (w_{sl}). Table 3.6 shows a summary of the measured dimensions of the key design parameters. Based on these values, the simulation model of the EM-coupled SL-to-GCPW signal transition is modified and compared with the initial simulation and measurement results [21].

Parameter (see Fig. 3.14d)	Simulated (µm)	Manufactured left (µm)	Manufactured right (µm)
δy_1	212	246	166
δy_2	198	132	242

Table 3.5: Layer misalignment analysis of EM-coupled SL-to-GCPW signal transition (YZ plane) [21] © 2018 IEEE

The probe-based S parameter measurement of a prototype of the EM-coupled SL-to-GCPW signal transition is carried out in the frequency range of 110 to 170 GHz. The measurement is done by using GSG probes of 150 µm pitch. A TRL calibration is performed before carrying out the measurements. For both RC and TC, the measured results are compared with the initial simulation result as well as the modified simulation result, as shown in Fig. 3.15. The two-port RCs, namely, S_{11} and S_{22} are shown in Fig. 3.15a and Fig. 3.15b, respectively. The resonances observed in the measured S_{11} curve (Meas. S_{11}), i.e., at around 115 GHz and 148 GHz, coincide with the resonances observed in the modified simulation S_{11} curve [Sim.(mod) S_{11}]. At around 137 GHz, a discrepancy is observed between the Meas. S_{11} and the Sim.(mod) S_{11} curve. On the other hand, the resonances observed in the measured S_{22} curve (Meas. S_{22}) are visible in the modified simulation S_{22} curve [Sim.(mod) S_{22}], but with a slight frequency shift. In addition, the Meas. S_{11} is less than −8.5 dB between 134 and 149 GHz (i.e., 10.6% relative bandwidth) and the Meas. S_{22} is less than −10 dB between 129 and 142 GHz (i.e., 9.6% relative bandwidth). Further, the Meas. S_{11} and the Meas. S_{22} curves show dissimilarity and the desired impedance matching is shifted to higher frequencies due to the XZ- and YZ-plane misalignment caused by the LTCC manufacturing defects. The measured, initial simulation and modified simulation TC curves are shown in Fig. 3.15c. The measured S_{21} curve (Meas. S_{21}) shows an excellent agreement with the modified simulation S_{21} curve [Sim.(mod) S_{21}]. For example, in the Meas. S_{21} curve, a sharp attenuation is observed between 150 and 160 GHz, which is clearly visible in the Sim.(mod) S_{21} curve as well. Moreover, this effect is not observed in the Sim.(init) S_{21} curve, which validates the correctness of the X-ray analysis. In the frequency range of 120.5 to 135 GHz, the back-to-back TC (i.e., Meas. S_{21}) is greater than −4 dB, which corresponds to a relative bandwidth of 11.35%. Therefore, for the single-ended EM-coupled SL-to-GCPW signal transition, the TC is greater than −2 dB in the desired frequency range [21].

Parameter (see Fig. 3.11)	manufactured left (μm)	manufactured right (μm)
w_{gcpw}	124 (max) 94 (min)	124 (max) 94 (min)
s_{gcpw}	32	32
l_1	166	85
g	98	98
l_{ap}	166	171
w_{ap}	491	493
l_3	50	161
w_{sl}	120	120

Table 3.6: Measured dimensions of EM-coupled SL-to-GCPW signal transition prototype [21] © 2018 IEEE

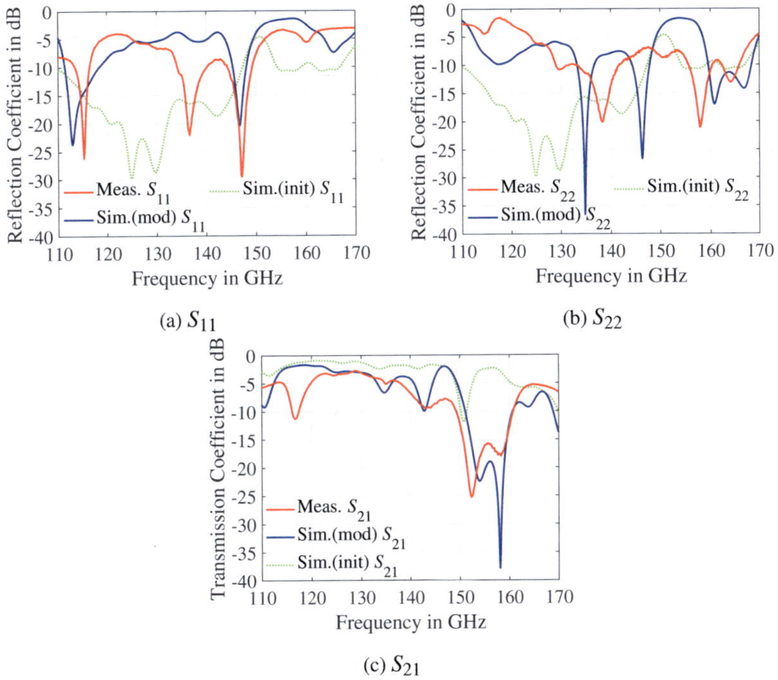

(a) S_{11}

(b) S_{22}

(c) S_{21}

Figure 3.15: Measured and simulated S parameters of EM-coupled SL-to-GCPW signal transition [21] © 2018 IEEE

3.3 Summary

In this chapter, two different SL-to GCPW signal transitions, namely, a via-based and an EM-coupled signal transition have been investigated. A via based signal transition operating from DC to 150 GHz is implemented by designing a controlled-impedance signal via. A signal via is first analyzed theoretically on the basis of its lumped element model. The approximate values of L_{via}, C_{via} and Z_{via} are calculated for a particular via design. This principle is further verified by means of a parameter study, performed on two via-based SL-to-GCPW signal transition models, namely, with air cavity (AC) and without air cavity (NC). The air cavities are included with the purpose of reducing the parasitic capacitance effect. The following via parameters: r_{via}, h_{via}, r_{cp} and d (see these parameters in Fig. 3.2) are varied and their influence on Z_{via} is analyzed on the basis of simulated impedance plots. Simultaneously, in each case, the RC and TC are simulated from DC to 170 GHz. Hence, an optimum design is achieved for the via-based signal transition. Multiple prototypes of both AC and NC models are manufactured. The top and buried MLs of the prototypes are analyzed by using X-ray microtomography, based on which the initial simulation models are modified. The measurement results of both (with and without air cavity) prototypes show an excellent agreement with their corresponding modified simulation results, thus validating the X-ray analysis. As per measurement, the via-based SL-to-GCPW signal transition with air cavity (AC) shows an RC less than −10 dB and a single-ended TC greater than −0.85 dB from DC to nearly 150 GHz. In contrast, the RC and the single-ended TC of the via-based SL-to-GCPW signal transition without air cavity (NC) is 2.5 dB higher and 0.35 dB lower, respectively, at 150 GHz. An EM-coupled signal transition is designed by means of an aperture, placed in the common ground plane of an SL and a GCPW. The coupling of EM fields is facilitated by using a modified OC structure. The modified OC consists of a GCPW with a large twin slot width and a signal conductor terminated at a certain distance from a semiannular ground ring. The initial simulation result of the signal transition shows an RC less than −14 dB, a single-ended TC greater than −1 dB and a radiation loss of approximately 10% in the frequency range of 114.5 to 132 GHz. Multiple prototypes of the signal transition are manufactured and the top and buried MLs of these prototypes are thoroughly investigated using X-ray microtomography. The most significant observation made while analyzing the manufacturing deviations is that the buried SL signal conductor is misaligned along the transversal and longitudinal directions by approximately 40 µm and 50 µm, respectively. Based on the X-ray analysis of the manufacturing deviations, the initial simulation model of the EM-coupled signal transition is modified. A good agreement is achieved between the measurement and the modified simulation results. As per the measurement, the EM-coupled signal transition shows a single-ended TC greater than -2 dB from 120.5 to 135 GHz. In addition, the measured two-port RCs are found to be shifted to higher frequencies due to the LTCC manufacturing deviations.

Due to the robust nature and excellent broadband performance of the via-based SL-to-GCPW signal transition, it is used in conjunction with various SL feed antennas in the next chapter and subsequently, in the implementation of a 122 GHz horizontal-configuration SiP. On the other hand, the EM-coupled SL-to-GCPW signal transition is found to be susceptible to the LTCC manufacturing deviations and is therefore not further dealt with in this work.

4 mm-Wave Antennas in LTCC Technology

A mm-wave SiP transceiver can transmit and receive mm-wave signals in an efficient way, provided its constituent antennas demonstrate an adequate RC, antenna gain and radiation efficiency over the operating frequency range of the MMIC. In addition, it is important to suppress the surface waves, which could otherwise severely degrade the radiation characteristics of the antennas. Taking these points into consideration, multiple variants of two different antenna concepts are implemented in the LTCC technology, namely, a via-fence (VF) and an EBG-based antenna concept. The feeding technique used for both concepts is an aperture-coupled feed. In this technique, the radiating element and the TL feeding the radiating element are placed on two vertically-stacked substrates. The coupling between the radiating element and the TL feed is achieved by means of an aperture placed in an intermediate ground plane separating the two substrates. This technique has the following advantages [GBBI01]:

- The thickness of the substrates supporting the radiating element and the TL feed can be chosen independently. The substrate supporting the radiating element should be relatively thick, in order to improve the radiation efficiency and the impedance bandwidth of an antenna (this is not necessarily true for an EBG antenna). In contrast, the TL feed should be designed on a relatively thin substrate so that the associated EM field is tightly bound to the substrate, thereby minimizing an undesired radiation and EM coupling.

- The radiating element is shielded from the spurious feed radiation due to the presence of an intermediate ground plane, separating the two substrates.
- It exploits the multilayer capability of the LTCC technology. Various configurations of planar TLs can be used to feed the radiating element, e.g., MS, GCPW and SL.

Two variants of an LTCC-based VF antenna with an MS feed, operating at 122 GHz, have been previously demonstrated in [BPR$^+$12] and [HBRZ13]. The VF antennas presented in section 4.1 include the following variants:

- VF stacked-patch (SP) antenna with an MS feed (Its performance is compared with the antennas shown in [BPR$^+$12] and [HBRZ13])
- VF SP antenna with an SL feed
- 2×2 VF patch array with an SL feed network

In section 4.2, variants of a 122 GHz Mushroom electromagnetic bandgap (MEBG) antenna are shown. An LTCC-based EBG antenna operating at a frequency above or even close to 100 GHz has not been demonstrated until now. Hence, a novel attempt has been made in this area. The following MEBG antenna variants have been investigated in this work:

- 4×4 MEBG antenna with an MS feed
- 6×6 MEBG antenna with an MS feed
- 4×4 MEBG antenna with an SL feed
- 6×6 MEBG antenna with an SL feed

In order to enable probe-based antenna measurement, the antennas fed by an SL (or an SL network) are combined with an SL-to-GCPW via-based signal transition, which is covered in detail in section 3.1. The antennas fed by an MS are combined with an MS-to-GCPW signal transition.

4.1 Aperture-Coupled Via-Fence Antennas

The impedance bandwidth and radiation efficiency of an antenna are usually enhanced by using a relatively thick antenna substrate, but this in turn, increases the surface wave power loss. Moreover, the undesirable surface waves distort the radiation pattern of an antenna and enhance the (undesired) mutual coupling in an antenna array. The antenna design challenges are further compounded, since the antenna has to be designed at a high frequency of 122 GHz. In section 4.1.1, a VF, SP antenna with an MS-to-GCPW signal transition feed is used to discuss the following topics. First, the suppression of surface waves using a VF. Second, the role of a VF and an SP in enhancing the impedance bandwidth and realized gain of the antenna. Third, the influence of LTCC manufacturing tolerances on the impedance bandwidth of the antenna.

4.1.1 Via-Fence, Stacked-Patch Antenna with an MS-to-GCPW Feed

The multilayered structure of a VF, SP antenna with an MS-to-GCPW feed is shown in Fig. 4.1 [17]. It consists of three layers of DuPont 9K7PV substrate and four layers of Au metal. An MS feed line, printed on the backside of the bottom substrate layer, i.e., ML 4, is used to feed the antenna. An MS-to-GCPW signal transition is provided on the same ML, in order to perform a probe-based antenna measurement. Two ground vias are used to connect the lateral ground planes of the 50 Ω GCPW to the ground plane on ML 3. The ground vias help in suppressing an undesired parasitic mode at the MS-to-GCPW signal transition discontinuity. A rectangular aperture is provided in the middle of an intermediate ground plane on ML 3, which couples the EM field from the MS feed line to the radiating elements of the antenna located on ML 1 and ML 2. The radiating elements include two vertically-stacked patches, namely, patch 1 and patch 2 on ML 1 and ML 2, respectively, and a VF structure. The VF structure consists of two square metal strips on ML 1, shorted to the intermediate ground plane by using rows of vias. The VF serves the following two purposes. First, it helps in suppressing the undesired surface waves and second, it helps in enhancing the antenna bandwidth. The theoretical analysis of an aperture-coupled MS feed antenna without a VF, is described in standard books on antenna design. A comprehensive account of methods, e.g., TL model and cavity model, is given in [GBBI01]. The antenna design is optimized by means of full-wave EM simulations carried out in the CST MWS software. The optimized antenna parameter values are given in Table 4.1 [17].

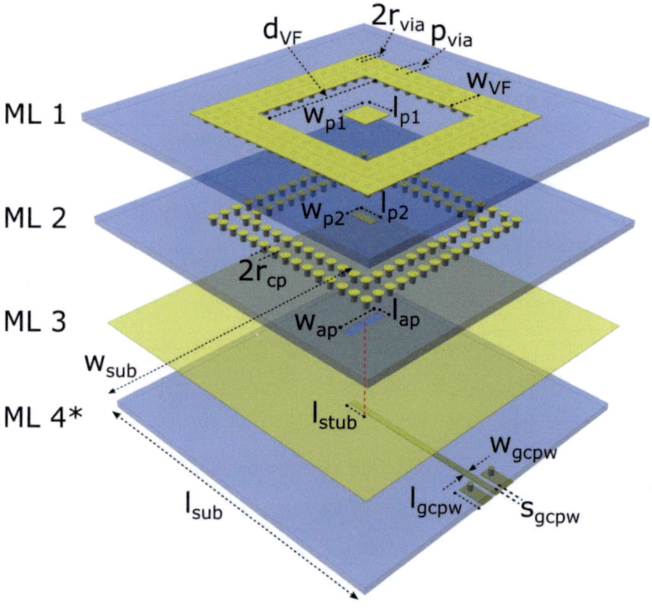

Figure 4.1: Multilayered structure of VF, SP antenna with MS-to-GCPW feed [17] © 2016 IEEE

Parameter (see Fig. 4.1)	Value (µm)	Parameter (see Fig. 4.1)	Value (µm)
l_{p1}	413	l_{p2}	345
w_{p1}	432	w_{p2}	143
r_{via}	41	l_{ap}	161
p_{via}	200	w_{ap}	707
w_{VF}	600	l_{stub}	219
d_{VF}	1900	w_{gcpw}	100
r_{cp}	72	s_{gcpw}	66
l_{gcpw}	400		

Table 4.1: Design parameter values of VF, SP antenna with MS-to-GCPW feed [17]

Suppression of Surface Waves and Radiation Pattern Improvement using a Via-Fence

The phenomena of surface wave excitation along an air-dielectric interface is shown in Fig. 4.2. These waves are excited at an elevation angle of θ, where $\arcsin(1/\sqrt{\varepsilon_r}) \leq \theta \leq \pi/2$, and undergo total reflection upon their incidence on the ground plane as well as on the air-dielectric interface. Therefore, these waves (largely) propagate inside the dielectric substrate in a zigzag manner, until they encounter a surface discontinuity. Thereafter, the surface waves undergo diffraction at the edges of a dielectric substrate and are therefore responsible for distorting the radiation pattern of an MS antenna. In addition, these waves enhance the undesired mutual coupling between the antennas in an array configuration [GBBI01]. In a grounded dielectric substrate, these waves propagate as transverse magnetic (TM) and transverse electric (TE) modes. The cutoff frequency (f_{c}) of these modes is given by Eq. 4.1.

$$f_{\mathrm{c}} = \frac{nc_0}{4h_{\mathrm{sub}}\sqrt{\varepsilon_r - 1}} \tag{4.1}$$

In the above equation, $n = 0, 2, 4...$ for TM_n modes and $n = 1, 3, 5...$ for TE_n modes. Since, TM_0 mode has 0 cutoff frequency, it is always excited in an MS antenna. The cutoff frequency for the next order surface wave mode, i.e., TE_1, is calculated for three different aperture-coupled VF antennas, as shown in Table 4.2.

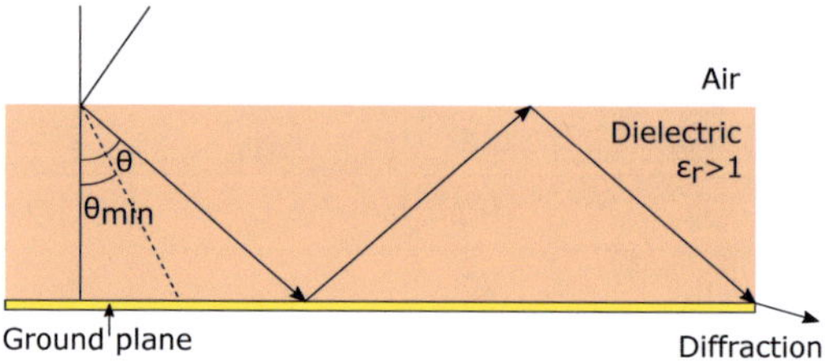

Figure 4.2: Surface wave in a grounded dielectric slab [GBBI01]

Antenna Type	$f_c(TE_1)$ (GHz)	h_{sub} (μm)	ε_r
This work	145.8	210	7.0
[BPR+12]	112.9	300	5.9
[HBRZ13]	101.2	300	7.1

Table 4.2: Influence of dielectric parameters on the TE_1 mode cutoff frequency

In Table 4.2, it is observed that in comparison to the previous two 122 GHz LTCC antennas, the aperture-coupled VF antennas shown in this work have a higher cutoff frequency for the TE_1 mode. Moreover, this cutoff frequency is much higher than the upper frequency limit of the MMIC selected in this work, i.e., 125.8 GHz. Hence, in this case, only the TM_0 mode needs to be suppressed. Further, it is important to note that the surface wave power loss can be neglected, provided the antenna substrate thickness satisfies the condition shown in Eq. 4.2 [GBBI01].

$$\frac{h_{\text{sub}}}{\lambda_0} \leq \frac{0.3}{2\pi\sqrt{\varepsilon_r}} \tag{4.2}$$

Using Eq. 4.2, it is estimated that an antenna substrate ($\varepsilon_r = 7.0$) should be 44 µm thick, in order to have a negligible surface wave power loss at 122 GHz. This condition cannot be satisfied using DuPont 9K7 substrate, since the minimum fired substrate thickness is approximately 105 µm. In fact, the minimum substrate thickness of other high frequency LTCC substrates, e.g., Ferro A6M, is also close to 100 µm. In addition, if the antenna substrate is too thin, it could lead to warpage during the LTCC co-firing process. To summarize, surface waves cannot be avoided altogether, hence their suppression becomes important for an efficient mm-wave antenna design. Some of the techniques used for this purpose are as follows:

- EBG: They are periodic structures, which do not allow EM waves to propagate within one or more specific frequency bands, referred to as "Bandgaps". The start and stop frequencies of these bandgaps depend on the geometrical dimensions and the substrate parameters used for forming an EBG unit cell. This approach is discussed in section 4.2.

- Embedded air cavity: An air cavity can be embedded in the antenna substrate, which lowers the effective dielectric constant of the antenna substrate. As a result, the surface wave power loss is reduced to a large extent. The process of embedding an air cavity in the buried substrate layers of an LTCC module requires a special manufacturing technique. Moreover, the probability of deformation of an embedded air cavity during the LTCC manufacturing process is quite high. Therefore, this approach is not used in this work.

- Composite substrate-superstrate structure: If the parameters of a substrate (on which the antenna is located) and a superstrate, namely, relative permittivity (ε_{r}), relative permeability (μ_{r}), substrate and superstrate thickness, satisfy a specific set of conditions, then a resonance is generated in the composite substrate-superstrate structure, such that the the TM_0 mode is not excited [AJ84]. The substrate-superstrate resonance condition is achieved if the superstrate has either a much higher ε_r or a much higher μ_r than the antenna substrate. In addition, the substrate and superstrate height should be in a specific ratio. The problems associated in implementing this approach in the LTCC technology are as follows. First, although dielectric inks with a very high value of ε_r are available in LTCC, their co-firing behavior with DuPont 9K7 substrate is not known. Second, materials with a high value of μ_r are not available in LTCC. Third, the LTCC substrates undergo shrinkage during the co-firing process and hence the substrate and superstrate height are susceptible to variation during the LTCC manufacturing process. Moreover, this approach improves the radiation efficiency only over a narrow bandwidth [AJ84].

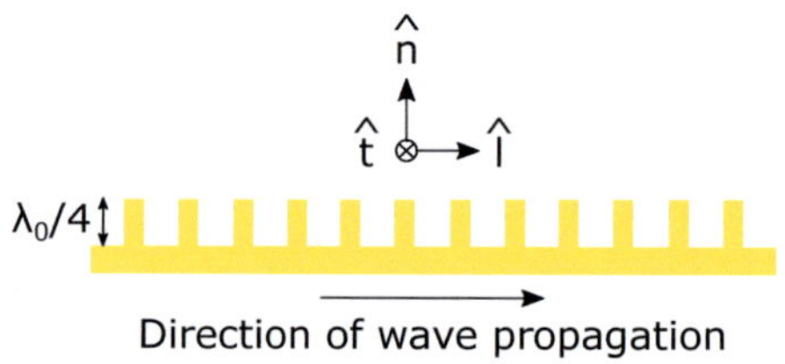

Figure 4.3: A conventional soft surface [Kil90]

- Soft surface: A conventional soft surface is shown in Fig. 4.3. It consists of an ideal conductor with corrugations transversal to the direction of wave propagation. The depth of these corrugations is $\lambda_0/4$. In H-plane, the soft surface acts as a perfect electric conductor (PEC). Since a PEC does not support the tangential component of an E-field, a TE mode cannot propagate. In E-plane, the transversal corrugations act as a shorted TL for the incident wave. This short circuit (SC) is transformed into an

OC at the aperture of the corrugated surface, since the corrugation depth is $\lambda_0/4$. Therefore, the soft surface has an infinite surface impedance in E-plane. An infinite surface impedance implies that the magnetic field (H-field) is almost negligible (since, surface impedance is the ratio of the E- and H-field components, which are orthogonal to each other). Therefore, the soft surface does not support the tangential components of an H-field in E-plane, i.e., the soft surface acts as a perfect magnetic conductor (PMC) in E-plane and hence a TM mode cannot propagate. To summarize, a soft surface acts as a PEC in H-plane and as a PMC in E-plane, thus preventing both TE and TM surface wave modes from propogating [Kil88] [KKM05]. Mathematically, a soft surface is defined by the conditions shown in Eq. 4.3 [Kil88] [Kil90].

$$Z_\mathrm{l} = \frac{-E_\mathrm{l}}{H_\mathrm{t}} = \infty \tag{4.3a}$$

$$Z_\mathrm{t} = \frac{E_\mathrm{t}}{H_\mathrm{l}} = 0 \tag{4.3b}$$

A theoretical derivation of this equation is given in [Kil90]. In this equation, the parameters Z_l (Z_t), E_l (E_t) and H_l (H_t) denote the longitudinal (transversal) components of the surface impedance, E-field and H-field, respectively.

Besides a transverse corrugated conductor, a soft surface can be realized in several ways. For example, a grounded dielectric substrate with transverse metal strips on top or a grounded dielectric substrate with transverse metal strips connected to the ground using vias can also be used to realize a soft surface [KK03]. In this work, a VF, (i.e., a grounded dielectric substrate with metal strips connected to the ground by means of vias) is used, which is a modern realization of the soft surface concept. This structure has been previously reported in [LDT+03]. The height of the VF is approximately 210 µm, which is close to $\lambda_\mathrm{g}/4$ in DuPont 9K7 substrate at 122 GHz. The contribution of the VF in suppressing the surface waves is analyzed in the following manner. The VF, SP antenna shown in Fig. 4.1 is simulated with and without the VF. The simulation is carried out in the CST MWS frequency solver. The absolute magnitude of the E-field distribution on the top surface of the antennas with and without the VF are shown in Fig. 4.4a and Fig. 4.4b, respectively. On comparing these two plots, the E-field suppression due to the VF is clearly visible. It is observed that the E-field magnitude at the bottom-left corner of the antenna substrate, i.e., at $x = -2.5$ mm, $y = -2.5$ mm, is 74.2 dBV/m and 83.7 dBV/m with and without the VF, respectively. Therefore, the VF suppresses the E-field magnitude by 9.5 dB. The E-field magnitude at the remaining three corners of the antenna substrate are also compared and a similar result is obtained. Next, the farfield radiation patterns of the antenna at 122 GHz with and without the VF are observed. The H- and E-plane radiation patterns are shown in Fig. 4.5a and Fig. 4.5b, respectively. The key radiation pattern parameters of both antenna variants are shown in Table 4.3. It is observed that the VF increases the boresight antenna gain by 2.8 dB, decreases the side lobe level by 3.1 dB in H-plane and 8.4 dB in E-plane, and decreases the back lobe level in both H- and E-plane by approximately 8.5 dB.

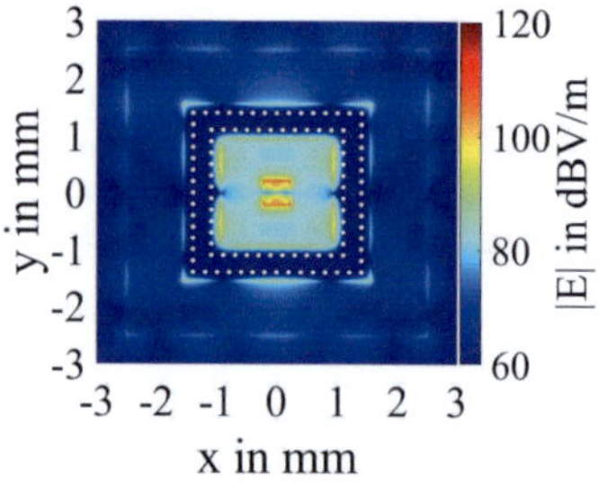

(a) E-field magnitude with VF

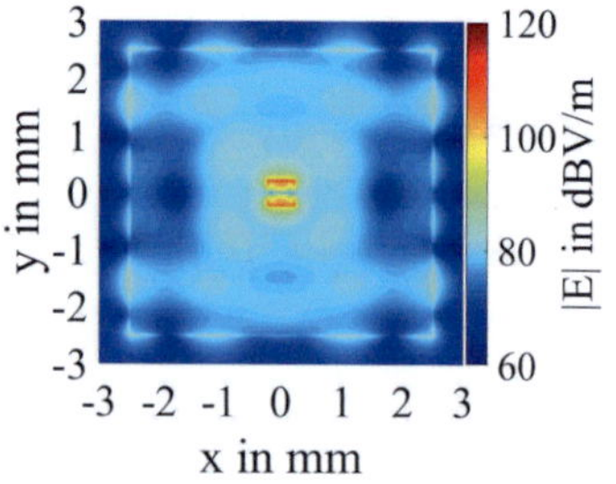

(b) E-field magnitude without VF

Figure 4.4: Simulated E-field with and without VF at 122 GHz

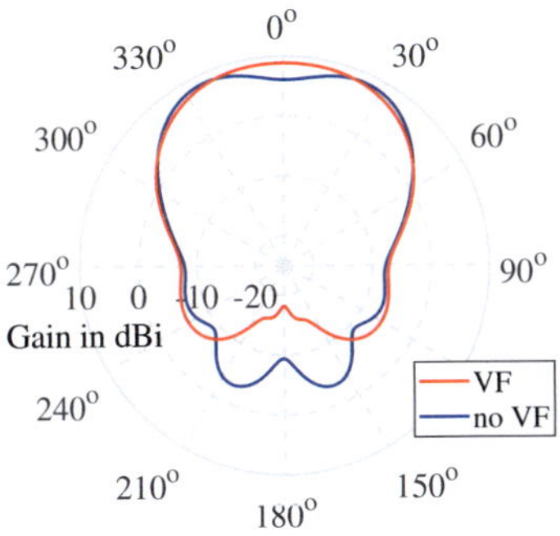

(a) H-plane

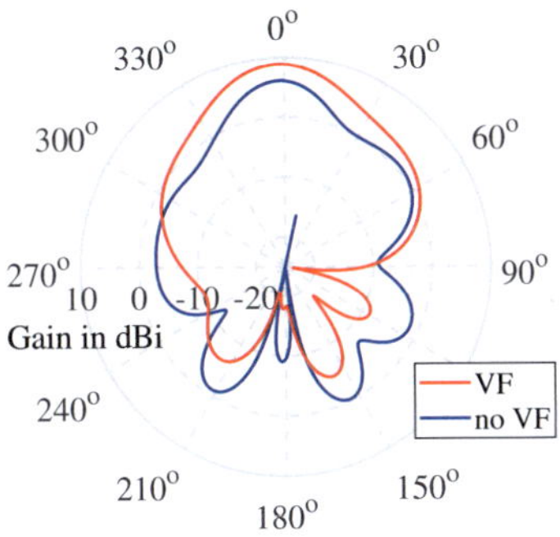

(b) E-plane

Figure 4.5: Simulated farfield radiation pattern with and without VF at 122 GHz

Bandwidth Enhancement

The impedance bandwidth of the aperture-coupled VF, SP antenna, shown in Fig. 4.1, is enhanced by means of coupled resonances. The physical features of the antenna contributing to the coupled resonances include the ground plane rectangular aperture, vertical SPs and VF. These resonances are seen as loops in the Smith chart. The size of the loop indicates the coupling strength. A small loop indicates a lower coupling strength and a large loop indicates a higher coupling strength [TWP98]. In order to enhance the antenna bandwidth, the location and spacing of the individual resonances should be adjusted so that both over-coupling and under-coupling of the resonances is avoided. The concept of using an SP for increasing the impedance bandwidth of an aperture-coupled antenna has been previously shown in [TWP98] and more recently in [LS11]. These research works dealt with frequencies below 100 GHz and a VF structure is not employed in either of them. The SP configuration has the following two advantages. First, in comparison to the coplanar parasitic patches, an SP does not increase the surface area of an antenna. Therefore, in an array configuration, the distance between the antennas can be kept under half-wavelength, thus avoiding the problem of grating lobes. Second, the radiation pattern and the phase center in an SP configuration remains symmetrical over the entire frequency band of operation, which is beneficial for an array.

Antenna Type	Boresight gain (dBi)	Side lobe H/E-plane (dB)	Back lobe H/E-plane (dB)
VF	8.8	-14.7/-13.7	-17.8/-18.1
No VF	6.0	-11.6/-5.3	-9.3/-9.5

Table 4.3: A comparison of the simulated farfield radiation pattern parameters with and without VF at 122 GHz

The contribution of the VF and SP in enhancing the impedance bandwidth of an antenna is analyzed by simulating the RC of the antenna (shown in Fig. 4.1) under the following three conditions.

1 Using the patch on ML 1 and the VF as resonators (the VF structure is used, but the SP configuration is not used). This test case is denoted as 'VF (no SP)' in the simulation plots shown in Fig. 4.6.

2 Using the patches on ML 1 and ML 2 as resonators (the SP configuration is used, but the VF structure is not used). This test case is denoted as 'SP (no VF)' in the simulation plots shown in Fig. 4.6.

3 Using the patch on ML 1, the patch on ML 2 and the VF as resonators (the VF structure as well as the SP configuration are used). This test case is denoted as 'VF + SP' in the simulation plots shown in Fig. 4.6.

The following observations are made from the Fig. 4.6a. If only the VF structure is used [i.e., VF (no SP)], the RC is less than -10 dB from 110.12 to 118.64 GHz, which corresponds to a relative bandwidth of 7.45%. If only the SP configuration is used [i.e., SP (no VF)], the RC is less than -10 dB from

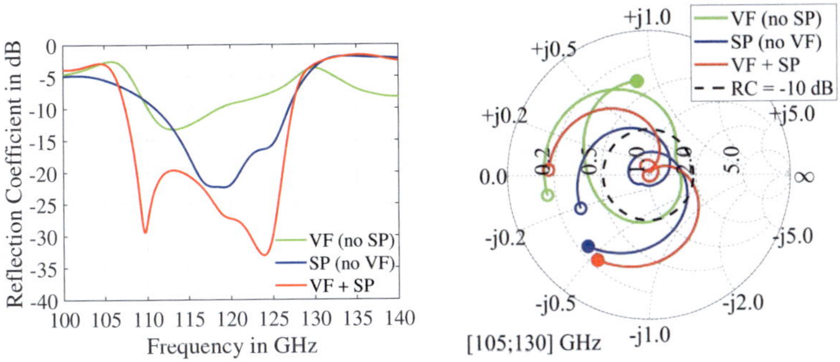

(a) RCs in dB

(b) RCs on Smith chart

Figure 4.6: Simulated RC of antennas with VF only [VF(no SP)], SP only [SP(no VF)], VF and SP [VF + SP]

111.04 to 126.92 GHz, which corresponds to a relative bandwidth of 13.35%. Finally, if the VF structure as well as the SP configuration are used (i.e., VF + SP), the RC is less than -10 dB from 107.48 to 127.36 GHz, which corresponds to a relative bandwidth of 16.93%. Therefore, the largest impedance bandwidth is achieved when the VF structure as well as the SP configuration are used. Fig. 4.6b shows the Smith chart representation of the RC in the above mentioned cases. As mentioned before, the resonances are seen as loops in the Smith chart. The RC curve for the 'VF + SP' case shows two tight loops, which indicate that the resonances due to the patches on ML 1 and ML 2 and the VF are adequately coupled. The RC curve for the 'SP (no VF)' case shows only one big loop, which indicates the absence of a resonance due to the VF structure. Finally, the RC curve of the 'VF (no SP)' case shows the biggest loop, which indicates a weak coupling between the resonances of the patch on ML 1 and the VF structure. Note that the distance between the patch on ML 1 and the VF structure is quite large, which results in a weak coupling. The coupling between these two elements can be increased by reducing the gap between them, but in that case, the patch on ML 2 loses its purpose, as it does not provide a benefit in terms of the impedance bandwidth and gain of the antenna.

Influence of LTCC Manufacturing Tolerances

It is important to verify if an antenna design is robust enough to withstand the standard LTCC manufacturing tolerances. Therefore, the influence of varying six key parameters of the antenna design (shown in Fig. 4.1) is investigated, namely, l_{p1}, l_{p2}, r_{via}, h_{sub}, l_{ap} and w_{ap}. In each case, only the parameter under test is varied in steps of 20 µm or 40 µm and the remaining parameters are kept constant. The influence of varying the parameter under test is analyzed by observing the impedance locus on the Smith chart. The results corresponding to these parameters are shown in Fig. 4.7 [17]. For each Smith chart representation, the reference impedance is 50 Ω and a reference circle with an RC of −10 dB is used to compare the impedance locus in each case. Consequently, it is checked if a parameter variation leads to a large spreading of the impedance locus, which in turn implies a severe reduction in the impedance bandwidth of the antenna [17].

- Substrate height (see Fig. 4.7a): The antenna performance shows the highest senstivity towards variation in the substrate height, h_{sub}. The parameter h_{sub} is varied in three steps namely, 95 µm, 105 µm and 115 µm. As h_{sub} increases, the size of both loops in the impedance loci decreases, which indicates an overall reduction in the coupling strength. Consequently, the impedance bandwidth of the antenna decreases [17].
- Via radius (see Fig. 4.7b): As the via radius, r_{via}, is increased from 31 µm to 41 µm, the size of both loops in the impedance loci increases, which indicates an overall increase in the EM coupling strength. For $r_{via} = 51$ µm, the two loops combine together to form a mutual resonance, which causes a reduction in the antenna bandwidth. This confirms that the VF structure, besides suppressing the surface waves, is

also responsible for enhancing the impedance bandwidth of the antenna [17].

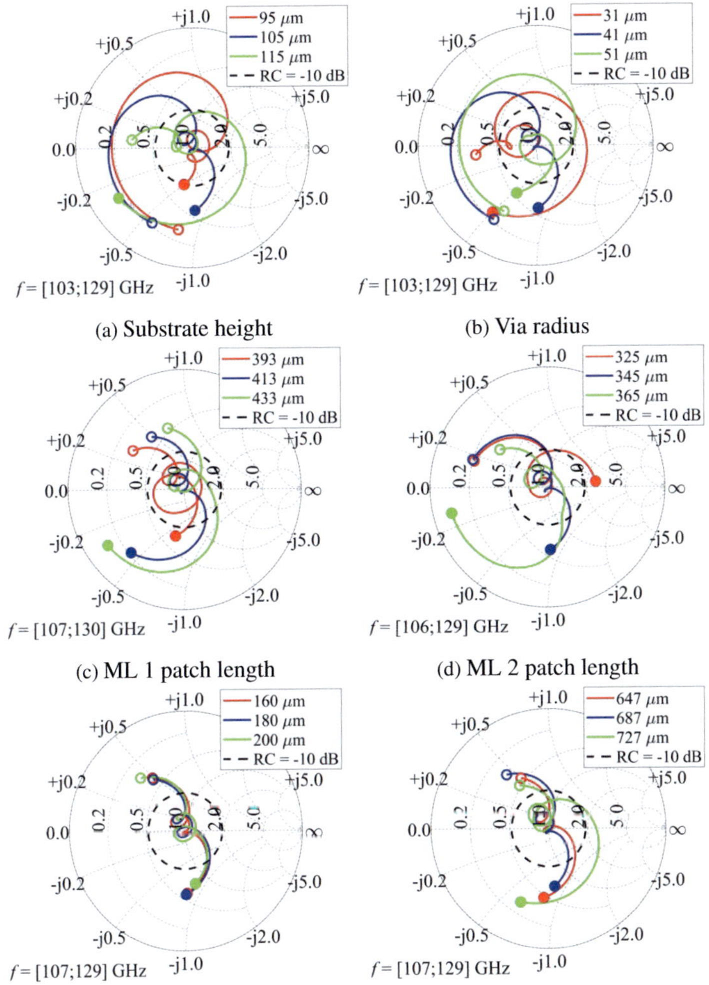

(a) Substrate height

(b) Via radius

(c) ML 1 patch length

(d) ML 2 patch length

(e) Ground plane aperture length

(f) Ground plane aperture width

Figure 4.7: Simulated RC of VF, SP antenna with MS-to-GCPW feed for various design parameter values [17] © 2016 IEEE

- Length of patches on ML 1 and ML 2 (see Fig. 4.7c and Fig. 4.7d): On observing the impedance locus for the ML 1 patch length, $l_{p1} = 393\,\mu m$, two loops corresponding to the lower and higher frequency resonances are clearly visible. As l_{p1} increases, the size of the upper loop decreases and shifts to a lower frequency, whereas the lower loop remains practically unaffected. This indicates that the higher frequency resonance results from the EM coupling between the patches on ML 1 and ML 2. As l_{p1} increases, it reduces the coupling between the two patches. On the other hand, as the length of the patch on ML 2, l_{p2}, is varied, the position of both loops in the impedance loci, is significantly altered. In other words, a variation in the parameter l_{p2} has a greater influence on the impedance bandwidth of the antenna, since the patch on ML 2 is EM coupled to the ground plane aperture as well as the patch on ML 1. In order to achieve the desired bandwidth enhancement, the l_{p2}/l_{p1} ratio should be suitably adjusted. For the chosen antenna substrate, the optimum value of l_{p2}/l_{p1} is found to be 0.84 [17].

- Ground plane aperture length and width (see Fig. 4.7e and Fig. 4.7f): As the ground plane aperture length, l_{ap}, is increased from 160 µm to 200 µm, a minor change is observed in the size of the loops in the impedance loci. Similarly, the ground plane aperture width, w_{ap}, is varied in bigger steps of 40 µm, in order to observe a significant change in the impedance loci. If the ground plane aperture is too wide, then a mutual resonance is observed, which leads to a reduction in the impedance bandwidth of the antenna [17].

Prototype and Measurement Result

Microscope images of the top and bottom layers of a prototype of the VF, SP antenna with an MS-to-GCPW feed are shown in Fig. 4.8a and Fig. 4.8b, respectively. In order to avoid deformation (i.e., warpage) of the antenna prototype during the LTCC manufacturing, the Au metal percentage in the antenna prototype had to be reduced. This was done by modifying the intermediate ground plane on ML 3. The intermediate ground plane is manufactured as a fine Au metal mesh in the area enclosed by the VF structure and the MS-to-GCPW signal transition feed. In the remaining substrate area, the intermediate ground plane on ML 3 is manufactured as a coarse Au metal mesh. An X-ray image of the antenna prototype is shown in Fig. 4.8c, in which the coarse metal mesh is clearly visible. The initial simulation model of the antenna shown in Fig. 4.1 is modified by replacing the solid ground plane on ML 3 with a ground plane mesh similar to the one seen in the X-ray image.

(a) Microscope photo (top view)

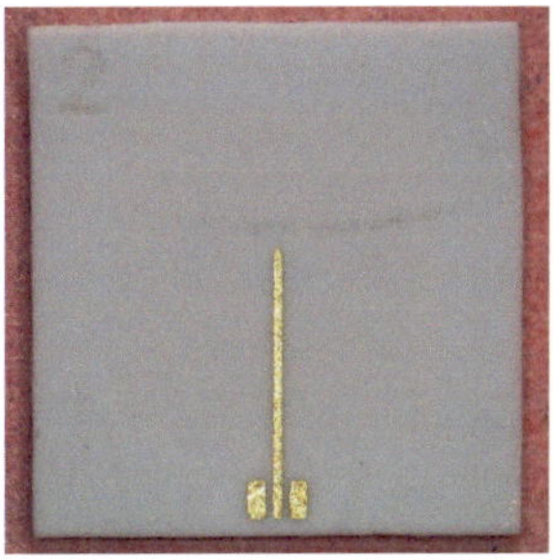

(b) Microscope photo (bottom view)

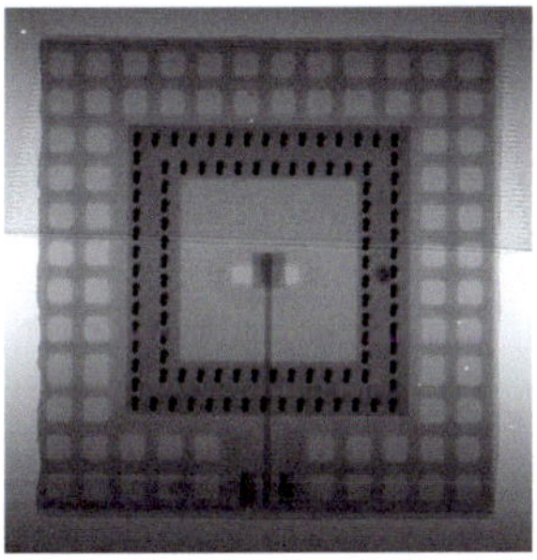

(c) X-ray image

Figure 4.8: A prototype of VF, SP antenna with MS-to-GCPW signal transition feed [17]

The antenna measurement results are compared with both the initial and modified simulation results. The prototype is measured using a probe-based antenna measurement setup shown in Fig. 4.9. The setup is described in [Bee13]. The same setup is used to measure all the antennas shown in this work.

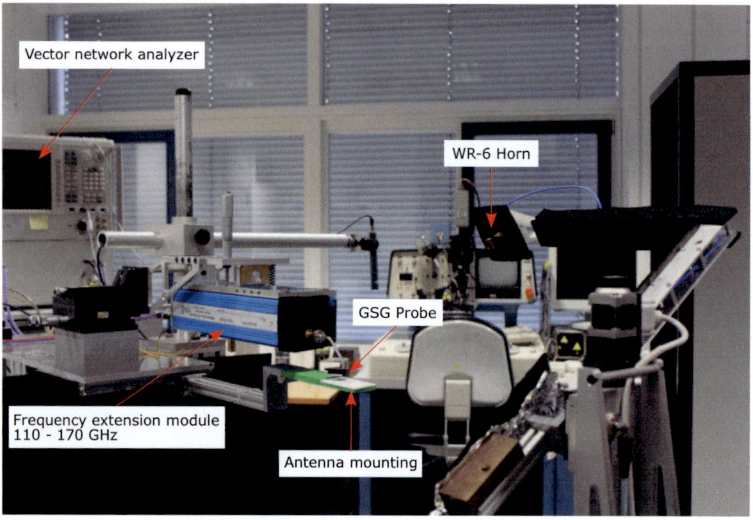

Figure 4.9: 110 to 170 GHz Probe-based antenna measurement setup

The measurement is carried out between 110 and 140 GHz, using a GSG probe of 150 μm pitch. The antenna measurement is preceeded by a gain calibration, which is done with a standard WR-6 horn antenna and a one-port short-open-load (SOL) calibration, which is done with a standard calibration substrate. The antenna radiates in a direction opposite to the 50 Ω GCPW pads, which are contacted by the GSG probe tips. The antenna is placed on a Rohacell ($\varepsilon_r \approx 1$) sheet in order to achieve a good approximation of the farfield boundary condition during the antenna measurement. The measured, initial simulation and modified simulation results are shown in Fig. 4.10.

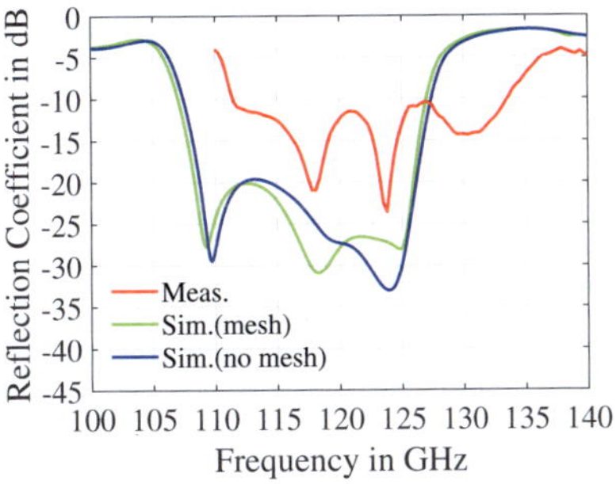

(a) RC

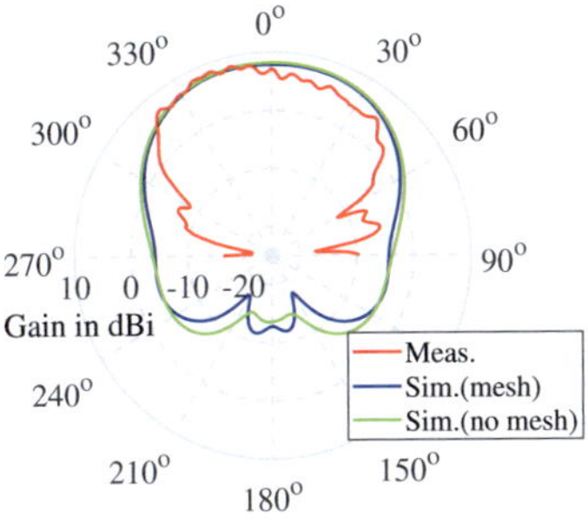

(b) Farfield radiation pattern in H-plane at 120 GHz

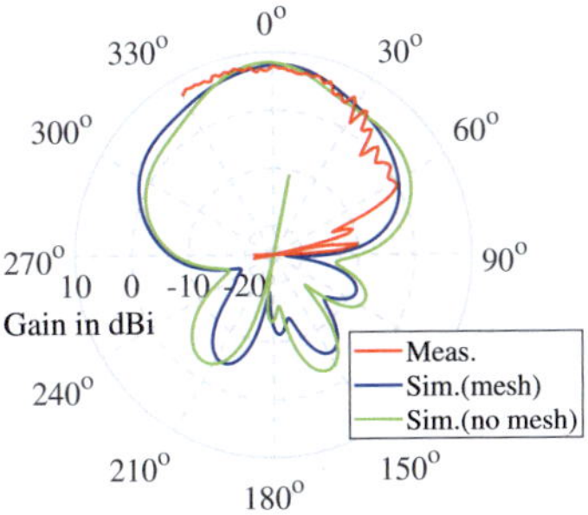

(c) Farfield radiation pattern in E-plane at 120 GHz

Figure 4.10: Measurement and simulation results of VF, SP antenna with MS-to-GCPW signal transition feed [17]

It is observed that the simulated RC of the antenna with a solid ground plane [i.e., Sim.(no mesh) RC] and with a mesh ground plane [i.e., Sim.(mesh) RC] are nearly the same. Therefore, the impedance matching of the antenna remains practically unaffected by the ground mesh. In the measured RC curve (i.e., Meas. RC), sharp resonances are observed at around 118, 124 and 130 GHz, which verifies the principle of bandwidth enhancement through coupled resonances. In comparison with the simulated RC curves, the measured RC curve is observed to be slightly shifted to a higher frequency range. This effect could be attributed to a variation of the substrate height caused by the substrate shrinkage, which occurs during the LTCC co-firing process. The measured RC is less than −10 dB between 111.5 and 133.5 GHz. Therefore, the measured impedance bandwidth of the antenna is 17.96%, which is close to the simulated impedance bandwidth of the antenna, i.e., 16.93%. The farfield radiation pattern of the antenna at 120 GHz in H- and E-plane are shown in Fig. 4.10b and Fig. 4.10c, respectively. The radiation pattern is measured over an angular span of 180° and 120° in H- and E-plane, respectively. The radiation pattern measurement in E-plane is restricted due to the mechanical construction of the antenna measurement setup [BZ10]. A good agreement is observed between the measured and simulated radiation patterns. The simulated antenna gain with the solid ground plane and the mesh ground plane is 8.2 dBi and 7.8 dBi, respectively at 120 GHz along the boresight direction. The corresponding value in measurement is 7.1 dBi. In addition, at 122 GHz, the boresight antenna gain is observed to be 8.8 dBi in the initial simulation with the solid ground plane, 7.7 dBi in the modified simulation with the mesh ground plane and 6.2 dBi in the measurement. Overall, it is observed that the ground mesh causes up to 1 dB reduction in the antenna gain. The simulated antenna efficiency between 110 and 125 GHz is greater than 88% and 86% in the simulation with solid and mesh ground plane, respectively.

4.1.2 Via-Fence, Stacked-Patch Antenna with an SL-to-GCPW Feed

The principle of an aperture-coupled VF, SP antenna described in the previous section, is used to design an antenna variant, using an SL-to-GCPW signal transition feed. The multilayered structure of the antenna is shown in Fig. 4.11 [22]. It consists of five DuPont 9K7PV substrate layers and six Au metal layers. Similarly to the previous antenna, it includes a metal patch on ML 1, a metal patch on ML 2 and a VF structure, which are EM coupled to an asymmetric SL feed. The EM coupling takes place through a rectangular aperture located on an intermediate ground plane on ML 3. Unlike the previous antenna, in this case, the intermediate ground plane is manufacturable as a solid metal plane (and not as a metal mesh). This becomes possible because the antenna consists of a higher number of substrate layers and therefore the volume percentages of the metal and substrate are appropriate for the LTCC co-firing process. The signal conductor of the asymmetric SL is located on ML 4. The intermediate ground plane on ML 3 and the back metal plane on ML 6 act as the top and bottom ground planes of the asymmetric SL. These two ground planes are connected by two rows of ground vias, placed on either side of the asymmetric SL signal conductor. The ground vias help in suppressing a parasitic parallel plate mode. In order to enable a probe-based S parameter measurement of the antenna and for interconnecting the antenna to a chip with GSG pads, the asymmetric SL signal conductor (on ML 4) is connected to the signal conductor of a 50 Ω GCPW line (on ML 3) by means of a signal via [22]. The concept and performance of a via-based SL-to-GCPW signal transition is discussed extensively in section 3.1. From a production standpoint, in order to make the 50 Ω GCPW pads accessible for either contacting with a probe or interconnecting with a chip (by means of wirebond or flip-chip), a surface cavity in the top two substrate layers, located above the 50 Ω GCPW line, is required. The width and length of the surface cavity are given by $w_{\text{sub}} = 3.5$ mm and $l_{\text{sub1}} - l_{\text{sub2}} = 0.75$ mm. The antenna design parameters are shown in Fig. 4.11 and their values are given in Table 4.4 [22]. Since a surface cavity is placed right above the GCPW line, it becomes difficult to manufacture an air cavity beneath the signal via as well (local concentration of multiple cavities in an LTCC module could lead to deformation during the LTCC manufacturing process). Therefore, the via-based SL-to-GCPW signal transition is manufactured without an air cavity below the signal via. This leads to a slight degradation in the impedance matching of the antenna.

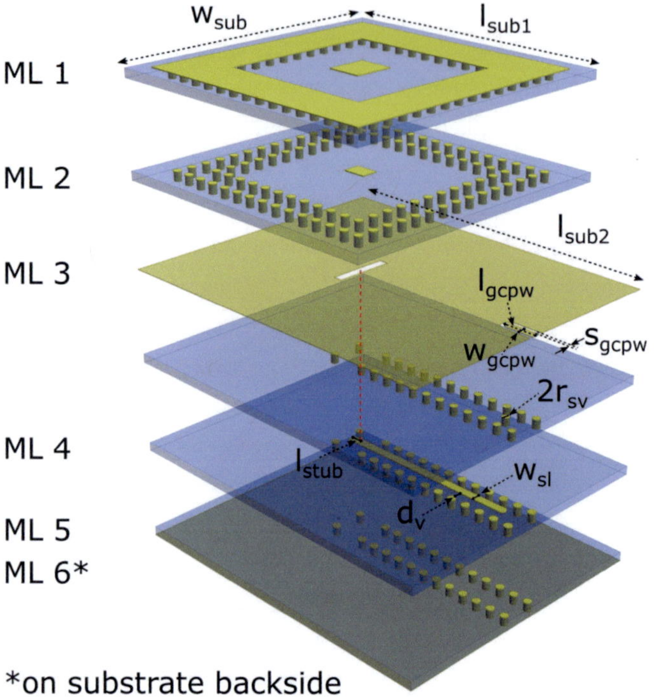

Figure 4.11: Multilayered structure of VF, SP antenna with SL-to-GCPW feed [22]

Prototype and Measurement Result

Multiple prototypes of the VF, SP antenna with an SL-to-GCPW feed are manufactured (the antenna prototype is shown in Fig. 5.2). The antennas are not diced and the adjacent samples are separated by a sufficiently large distance of around 50 mm, i.e., around $50\lambda_g$ in the desired frequency range. Therefore, the mutual coupling between the prototypes is practically negligible. The antenna measurement is performed using the same probe and calibration, as described in the previous section. The measurement and simulation results of the antenna without a cylindrical air cavity beneath the signal via are indicated as Meas.(NC) and Sim.(NC), respectively. The simulation result of the antenna

Parameter (see Fig. 4.11)	Value (μm)	Parameter (see Fig. 4.11)	Value (μm)
l_{p1}	421	l_{p2}	282
w_{p1}	482	w_{p2}	208
r_{via}	45	l_{ap}	117
p_{via}	200	w_{ap}	720
w_{VF}	600	l_{stub}	162
d_{VF}	1900	w_{gcpw}	100
r_{cp}	50	s_{gcpw}	41
l_{gcpw}	400	w_{sl}	100
r_{sv}	40	d_{v}	100
l_{sub1}	3500	w_{sub}	3500
l_{sub2}	4250		

Table 4.4: Design parameter values of VF, SP antenna with SL-to-GCPW feed [22]

with a cylindrical air cavity beneath the signal via (the air cavity dimensions are the same as shown in section 3.1), denoted as Sim.(AC) is also shown for comparison. The simulated and measured RCs are shown in Fig. 4.12a [22]. It is observed that the Meas.(NC) RC coincides with the Sim.(NC) RC from 110 to nearly 122 GHz. The Sim.(NC) RC shows a resonance at 135 GHz, which appears to be shifted to 130 GHz in the Meas.(NC) RC. Consequently, the Meas.(NC) RC is less than −10 dB between 116.9 and 132.2 GHz, thus achieving a relative bandwidth of 12.28%. In comparison to the Sim.(NC) RC, the Sim.(AC) RC, shows a larger impedance bandwidth and a 5 to 10 dB improvement in the RC magnitude. In terms of the realized antenna gain (see Fig. 4.12b [22]), the measured as well as the simulated gain curves show the peak antenna gain at 122 GHz. In addition, both measured and simulated gain curves show a small dip in the antenna gain at around 114 GHz and a sharp dip at around 132 GHz. The Meas.(NC) and the Sim.(NC) gain curves show an overall good agreement with each other. Since, the antenna substrate used in the measurement is much larger than in the simulation, it leads to some deviation in the gain values observed in the Meas.(NC) and Sim.(NC) curves. In comparison to the Sim.(NC) gain curve, the Sim.(AC) gain curve shows up to 1 dB higher antenna gain in the desired frequency range.

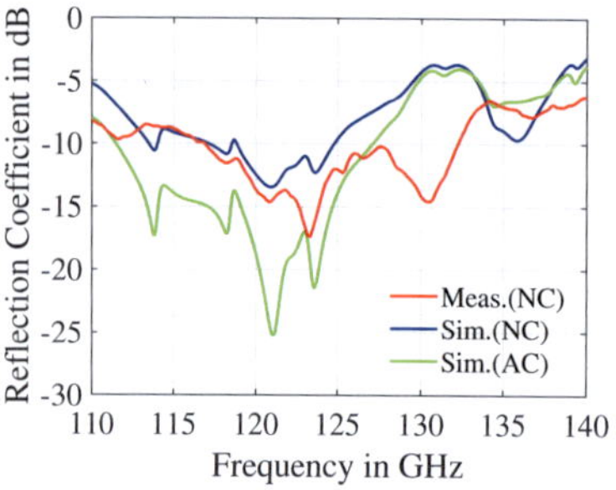

(a) RC

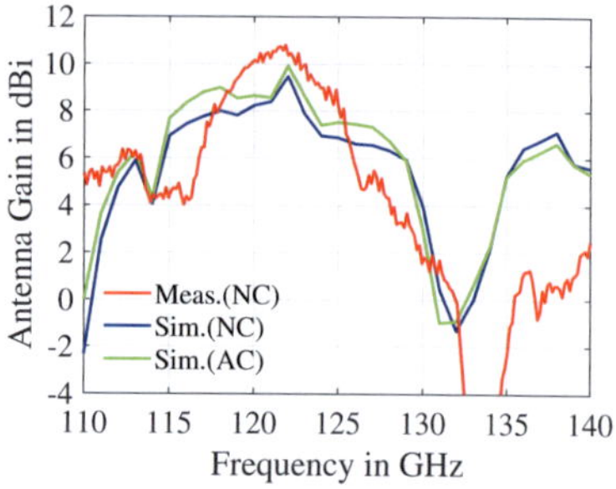

(b) Antenna gain

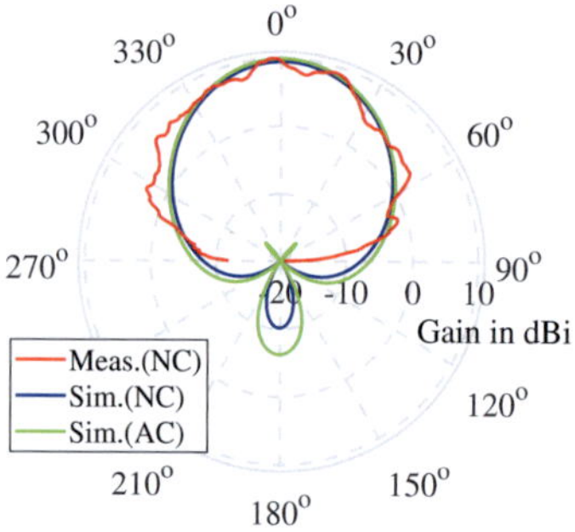

(c) Farfield radiation pattern in H-plane at 122 GHz

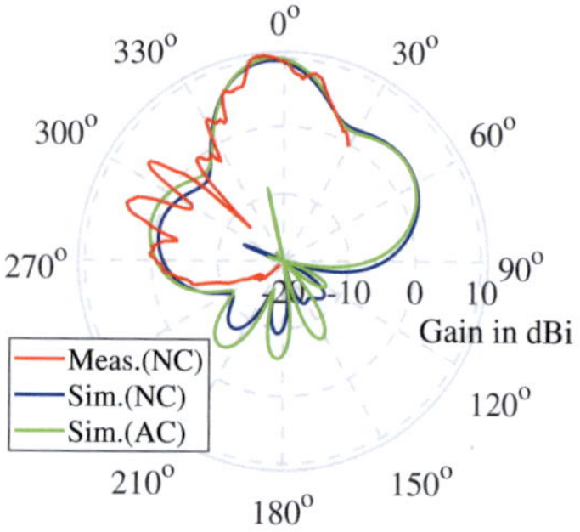

(d) Farfield radiation pattern in E-plane at 122 GHz

Figure 4.12: Measurement and simulation of VF, SP antenna with SL-to-GCPW feed [22]

The farfield radiation patterns at 122 GHz in H- and E-plane are shown in Fig. 4.12c and Fig. 4.12d, respectively [22]. The measured antenna gain along the boresight direction is 9.6 dBi and the measured peak antenna gain is 10.5 dBi, which occurs at a minor tilt of 6° from the boresight direction, in E-plane. The Meas.(NC) and the Sim.(NC) radiation patterns show a good agreement in both H- and E-planes, barring the enlargement observed at around 300° in E-plane, which occurs due to the influence of the measurement probe. Note that the farfield radiation pattern measured in the E-plane of the previous antenna with an MS-to-GCPW feed (see Fig. 4.10c) does not show this effect, since the antenna radiates in a direction opposite to the probe contacting the GCPW pads. The Sim.(AC) radiation pattern in H- and E-plane shows a higher back lobe radiation since the bottom air cavity is simulated with an 'open add space' boundary condition. In reality, the air cavity should be covered with a ground plane. The simulated efficiency of the antenna without an air cavity beneath the signal via is greater than 80% between 116 and 124.5 GHz.

4.1.3 2 x 2 Via-Fence Patch Array with an SL Network-to-GCPW Feed

An array is built using 2×2 antennas, where each antenna consists of a single patch surrounded by a VF. The multilayered structure of the array is shown in Fig. 4.13. The array is made using four layers of DuPont 9K7PV substrate and five layers of Au metal. In comparison with the previous two antennas, the distance between the patch and the VF is kept shorter, so as to reduce the size of each individual antenna and hence, decrease the distance between the array elements. The size of each antenna is 1.675 mm×1.675 mm and the center-to-center distance between any two adjacent antennas is 1.55 mm, i.e., $0.63\lambda_0$ at 122 GHz. The array elements, unlike the previous antennas, do not include a second metal patch on ML 2 (i.e., an SP configuration is not used in this case), because the coupling between the patch (on ML 1) and the VF is dominant in this case and therefore a vertically-stacked parasitic patch (i.e., an SP configuration) does not provide a benefit in terms of either the impedance bandwidth or gain of the array. The array elements are EM coupled through four rectangular apertures located on an intermediate ground plane (on ML 3) and the array feed is provided by an SL network. The signal trace, top and bottom ground plane of the SL network are located on ML 4, ML 3 and ML 5, respectively. The SL network consists of three quarter-wave matched T-junctions and four 90° curved bends. The SL signal conductor has a width of 100 µm, which provides an impedance of 35 Ω. A quarter-wave matched T-junction requires a stub of impedance $35\,\Omega/\sqrt{2} = 24.75\,\Omega$ and an electrical length of $\lambda_g/4$ at 122 GHz. These values are calculated using the LineCalc software from Keysight, which provides an initial SL stub width, $w_T = 200$ µm and a stub length, $l_T = 233$ µm. The values are optimized on the basis of EM simulations carried out in the CST MWS software. The optimized values are $w_T = 270$ µm and $l_T = 240$ µm. The SL network is surrounded by ground vias, which interconnect the top and bottom ground planes on ML 3 and ML 5, thereby avoiding the excitation of a parasitic parallel plate mode. A via-based signal transition is used to connect the main SL signal conductor to a 50 Ω GCPW line on ML 3. Note that the via-based signal transition connects a 35 Ω SL to a 50 Ω GCPW, whereas the via-based signal transition shown in the previous antenna and in section 3.1 connects a 41 Ω SL to a 50 Ω GCPW. Therefore, the via-based signal transition geometry is slightly modified in order to obtain an adequate RC and TC in the desired frequency range. The modified parameters of the signal transition are shown in Fig. 4.13. The optimized value of the key array design parameters are given in Table 4.5.

Via-Based Symmetric SL-to-GCPW Signal Transition

In comparison with the via-based signal transition shown previously, the following two modifications have been introduced in this signal transition variant. First, the via anti-pad radius r_{ap} is increased from 91 µm to 131 µm, which decreases the C_{via} and increases the Z_{via} of the signal via (see Eq. 3.4 and Eq. 3.5). As per the impedance plots shown in section 3.1.2, the Z_{via} associated with the signal via is close to 40 Ω. Second, a GCPW stub of width $w_1 = 180$ µm and a taper transition is introduced between the signal via catch-pad and the 50 Ω GCPW line. The GCPW stub provides an impedance of

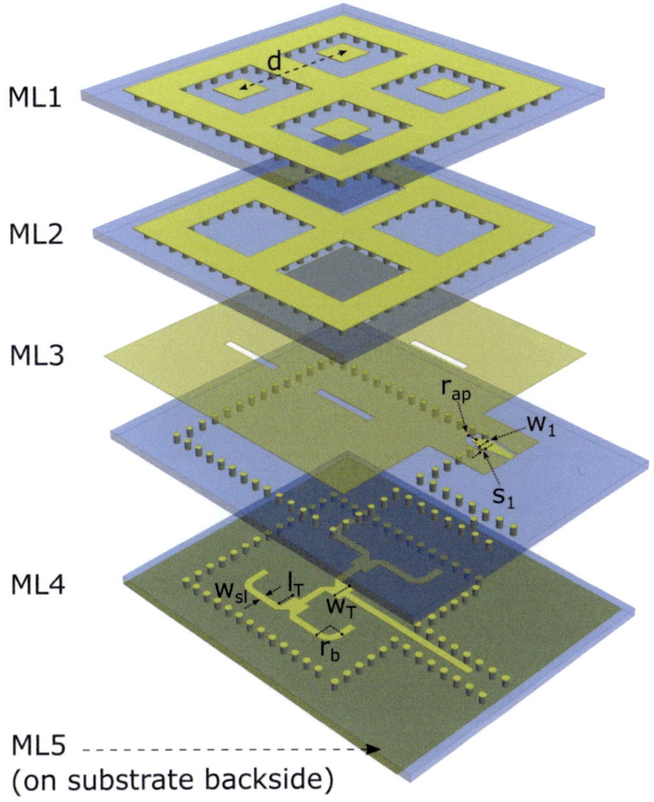

Figure 4.13: Multilayered structure of 2×2 VF patch array with SL Network-to-GCPW Feed

40 Ω, i.e., close to Z_{via}, and hence helps in impedance matching. A single-ended configuration of the signal transition is simulated using the CST MWS frequency solver and 'open add space' boundary conditions. Fig. 4.14a [10] shows the simulation model along with its waveguide ports (Port 1 faces the GCPW and Port 2 faces the symmetric SL). The simulated RC at the waveguide port 1 (S_{11}) and the simulated TC (S_{21} i.e., when the wave is incident on the waveguide port 1) are shown in Fig. 4.14b and Fig. 4.14c, respectively [10].

Parameter (see Fig. 4.13)	Value (µm)	Parameter (see Fig. 4.13)	Value (µm)
l_{p1}	425	l_{T}	240
w_{p1}	425	r_{b}	215
r_{via}	40	r_{ap}	131
p_{via}	200	w_1	180
w_{VF}	430	s_1	41
d_{VF}	1030	w_{gcpw}	100
l_{ap}	105	s_{gcpw}	41
w_{ap}	855	d_{v}	100
l_{stub}	115	w_{sub}	4000
w_{sl}	100	l_{sub1}	4000
w_{T}	270	l_{sub2}	5000

Table 4.5: Design parameter values of 2×2 VF patch array with SL Network-to-GCPW feed

In Fig. 4.14b, it is observed that the Sim. S_{11} is less than -10 dB up to a frequency of 158 GHz. In the operating frequency range of the antenna, the Sim. S_{11} is less than -15 dB. In terms of the TC (see Fig. 4.14c), the Sim. S_{21} is greater than -0.5 dB up to a frequency of 158 GHz. In contrast, the previous via-based signal transition based on three DuPont 9K7 substrate layers and an air cavity achieved a similar performance up to a frequency of 150 GHz. Therefore, it is once again verified that a higher number of substrate layers beneath a signal via leads to larger parasitic effects, which degrade the signal via performance at higher frequencies.

Estimation of Array Feed Network Loss

The loss incurred by the feed network of the antenna array is estimated in the following manner. The array elements, namely, Antenna 1, Antenna 2,

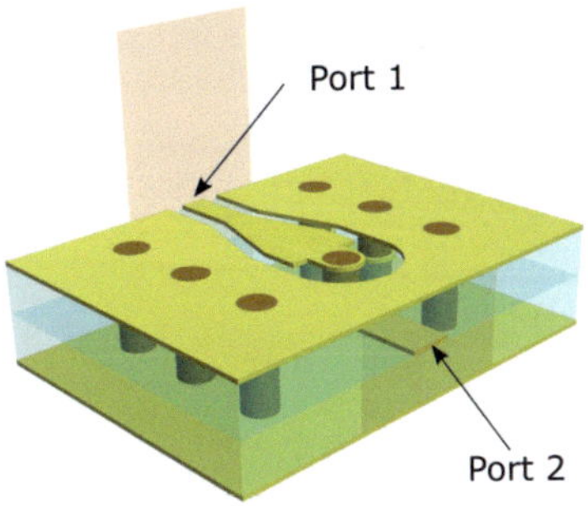

(a) Simulation model

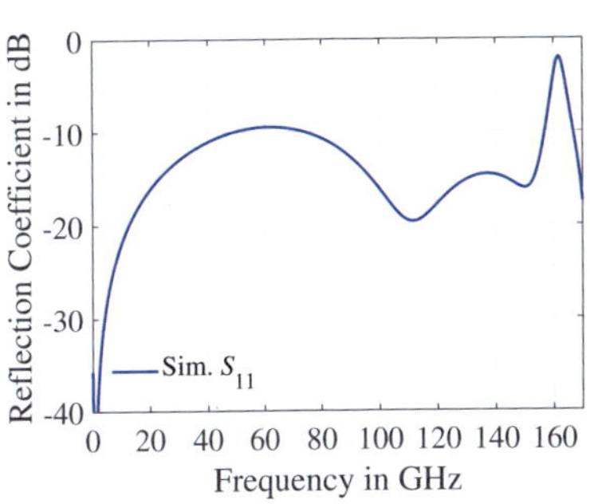

(b) Simulated S_{11}

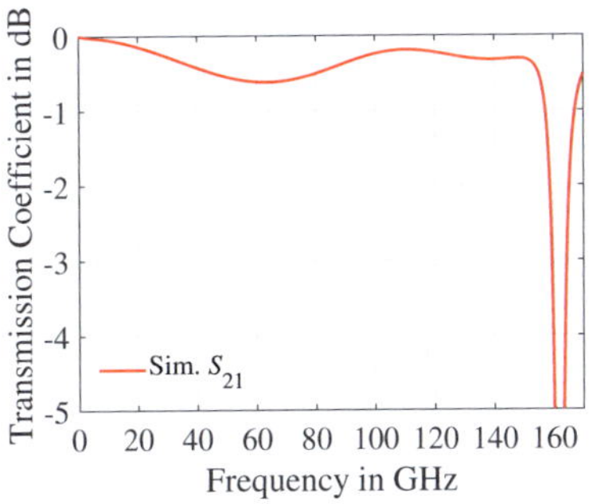

(c) Simulated S_{21}

Figure 4.14: Via-based symmetric SL-to-GCPW signal transition [10] © 2018 IEEE

Antenna 3 and Antenna 4, shown in Fig. 4.15a, are simulated individually using waveguide ports. The position of the waveguide ports is also shown in Fig. 4.15a. As seen in this figure, each array element is fed by a short SL stub of length 235 µm. Fig. 4.15c shows that in the frequency range of interest, the simulated RC of Antenna 1 (Sim. S_{11}), is less than -10 dB. Further, in the frequency range of interest, the simulated mutual coupling between Antenna 1 and Antenna 2 (Sim. S_{21}) is less than -15 dB, the simulated mutual coupling between Antenna 1 and Antenna 3 (Sim. S_{31}) is less than -21 dB and the simulated mutual coupling between Antenna 1 and Antenna 4 (Sim. S_{41}) is less than -25 dB. The resultant gains over frequency of the individual array elements are shown in Fig. 4.15d. It is observed that each array element shows nearly the same antenna gain in the frequency range of interest. At 122 GHz, the antenna gain achieved by Antenna 1, Antenna 2, Antenna 3 and Antenna 4 are 7.7 dBi, 7.3 dBi, 6.9 dBi and 6.6 dBi, respectively. Further, at 135 GHz, which lies outside the operating frequency range of the array, the gain of Antenna 3 and Antenna 4 show a sharp dip. This could be attributed to the asymmetrical positioning of the array elements on the substrate. Both Antenna 3 and Antenna 4 are located closer to the substrate edge and a dip in their antenna gains indicates the presence of surface waves, which leads to a distortion in their radiation patterns. The following two points should be noted here. First, if the antennas are located symmetrically on the substrate, then an additional bend becomes necessary in the main SL of the feed network, which will lead to an additional loss. Second, the VF structure suppresses the surface waves only within a finite bandwidth. Hence, a dip is seen in the gain of the individual antennas outside the frequency range of interest.

After simulating the individual antennas (or array elements), the schematic view in the CST MWS software is used to connect the waveguide ports (corresponding to Antenna 1, 2, 3 and 4) to three ideal 3 dB power dividers, as shown in Fig. 4.15b. The output ports of these 3 dB power dividers are terminated with an impedance of 35 Ω, as required by the SL stubs feeding the antennas. Thereafter, the simulated individual antenna gains are combined by applying post processing to the schematic model and hence, the ideal array gain is calculated. The ideal array gain is compared with the real array gain, which includes the SL network and the via-based symmetric SL-to-GCPW signal transition. The ideal as well as the real array gain over frequency curves are shown in Fig. 4.15d. It is observed that at 122 GHz, the difference between the ideal and real array gain is 1.8 dB. Since, the simulated TC (i.e., S_{21}) of the single-ended via-based symmetric SL-to-GCPW signal transition (see Fig. 4.14a and Fig. 4.14c) is 0.25 dB at 122 GHz, the loss incurred by the SL feed network is calculated as 1.55 dB, which is a reasonable value.

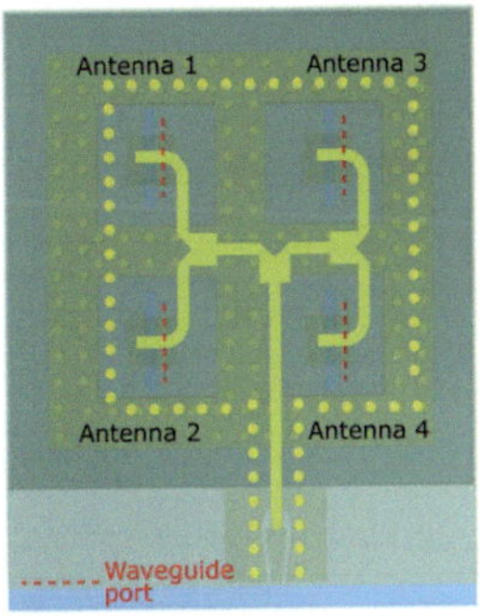

(a) Waveguide port position

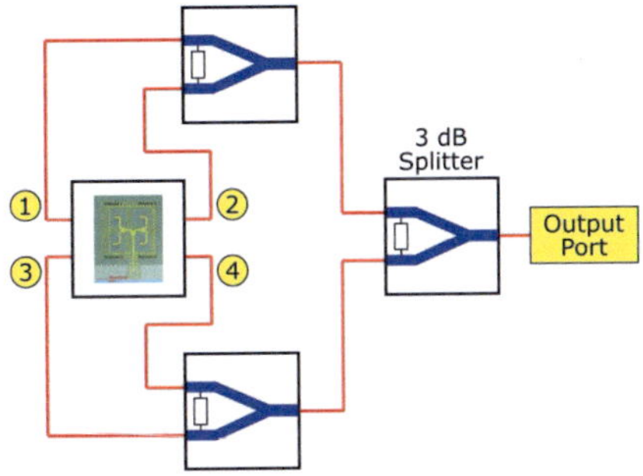

(b) Schematic model

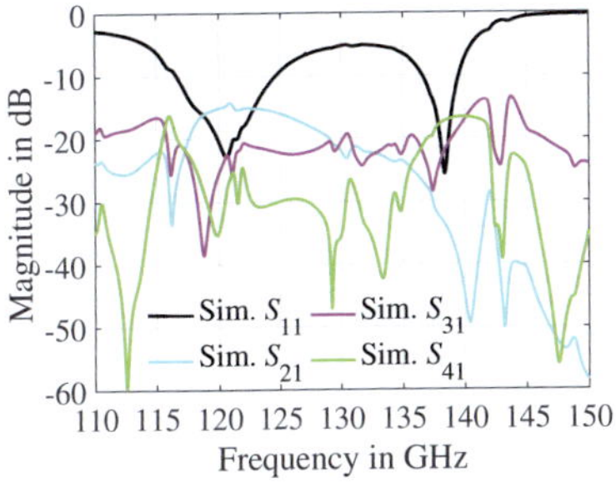

(c) Simulated S parameters of a single antenna (Antenna 1)

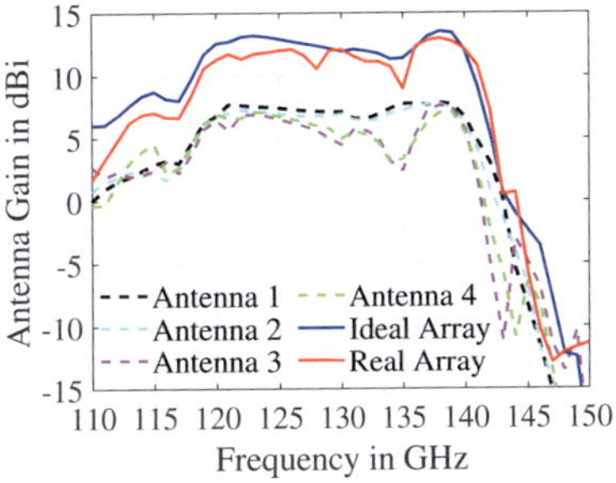

(d) Simulated antenna gain of array elements, array with an ideal feed network and array with the real feed network

Figure 4.15: Simulated feed network loss of 2×2 VF patch array with SL Network-to-GCPW feed

Prototype and Measurement Result

A prototype of the aperture-coupled 2×2 VF patch array with an SL Network-to-GCPW feed is shown in Fig. 4.16. The prototype is measured using the same probe and calibration, as specified in section 4.1.1. The measurement and simulation results of the array are shown in Fig. 4.17. The measured and simulated RCs shown in Fig. 4.17a are in good agreement with each other. The resonance frequencies observed in the measurement are seen in the simulation as well. A minor discrepancy is observed between the measurement and simulation, namely, a resonance seen at around 132 GHz in the measurement, seems to be shifted to a slightly higher frequency, when compared with the simulation. Consequently, the impedance bandwidth over which the measured RC is below −10 dB is slightly less than the simulation. The measured RC is less than −10 dB between 118 and 126.5 GHz, which results in a relative bandwidth of 6.95%. The measured and simulated array gain curves are shown in Fig. 4.17b. A comparison of the progression of the measured and simulated gain curves shows an overall good agreement and a minor shift over frequency is observed between them. The measured array gain is greater than 10 dBi between 119.3 and 129.5 GHz. The maximum array gain is observed as 12.85 dBi at 124.4 GHz. The measured and simulated farfield radiation patterns of the array at 122 GHz in H- and E-plane are shown in Fig. 4.17c and Fig. 4.17d, respectively. The measured radiation pattern tallies with the simulated radiation pattern in both planes. The array gain measured along the boresight direction is 11.3 dBi. The simulated efficiency of the array is greater than 70% between 117 and 130 GHz.

Figure 4.16: A prototype of 2×2 VF patch array with SL network-to-GCPW feed

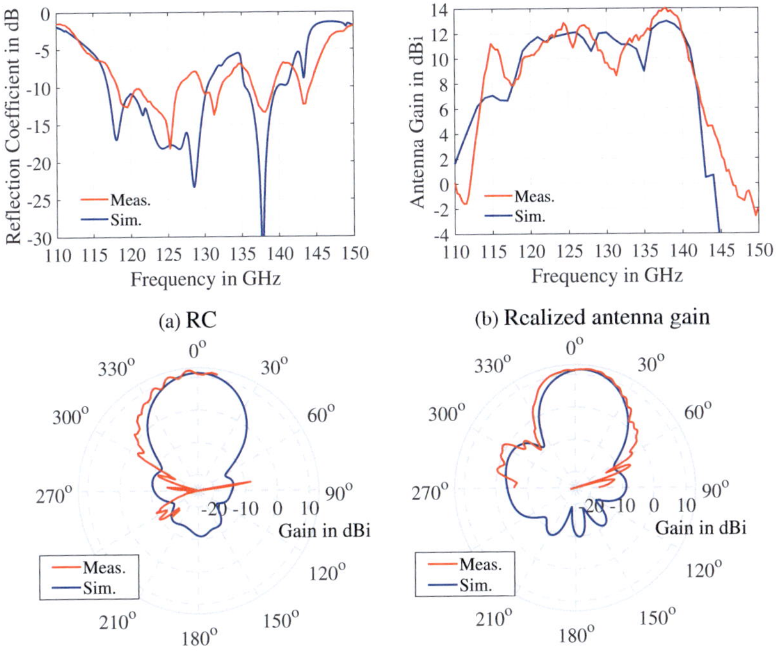

(a) RC

(b) Realized antenna gain

(c) Farfield radiation pattern in H-plane at 122 GHz

(d) Farfield radiation pattern in E-plane at 122 GHz

Figure 4.17: Measurement and simulation of 2×2 VF patch array with SL network-to-GCPW feed

4.2 Aperture-Coupled MEBG Antennas

In a typical patch antenna, the fundamental resonance occurs at a radial frequency (ω_0), at which the physical length of the patch is equal to half of the guided wavelength ($l_\mathrm{p} = \lambda_g/2$). Equivalently, the electrical length of the patch is given by $\beta l_\mathrm{p} = \pi$, where β is the phase constant. The higher-order resonances of the patch antenna occur at integral multiples of the fundamental resonance frequency, i.e., $\omega_n = n\omega_0$, where $n = 2,3,4...$ Consequently, these resonances are widely separated from one another, especially at high frequencies. This in turn, raises the question of how to increase the antenna bandwidth. The issue can be addressed using the following two approaches. In the first approach, additional resonating elements are used, whose resonance frequencies are located close to the fundamental resonance frequency of the main patch. This approach has been already covered in the previous section, in which a VF structure and an SP configuration are used to enhance the impedance bandwidth of the antenna. The second approach is based on an MEBG-based periodic structure, which forms the focus of this section. An MEBG is a two-dimensional EBG, which was first introduced in [SZB+99], under the name of a high impedance surface. In this section, four different configurations of an MEBG unit cell are used to realize 122 GHz LTCC antennas with a low profile, yet a high bandwidth-efficiency product.

4.2.1 Lumped Element Model of an MEBG Unit Cell

An MEBG unit cell consists of a dielectric substrate with a metal patch on its top surface, a solid ground plane on its bottom surface and a via connecting the metal patch and the ground plane in the center. The EM properties of an MEBG depend on the following design parameters: metal patch width w_{p}, gap between the adjacent metal patches g, via radius r_{via}, dielectric substrate height h_{sub} and ε_{r}. A 4×4 matrix of MEBG unit cells is shown in Fig. 4.18a. The behaviour of an MEBG can be understood from its lumped element equivalent circuit model, shown in Fig. 4.18b. The capacitance (C) and inductance (L) elements shown in this model are associated with the following physical effects: C_{L} arises due to the coupling between the adjacent metal patches, L_{R} arises due to the magnetic flux generated by the current flowing on the top metal patch, C_{R} arises due to the coupling between the top metal patch and the bottom ground plane and L_{L} arises due to the magnetic flux generated by the current flowing along the via [CI06] [YS09]. At low frequencies, L_{R} acts as an SC and C_{R} acts as an OC since, $X_{\mathrm{L}} = j\omega \mathrm{L}$ and $X_{\mathrm{C}} = 1/j\omega \mathrm{C}$, where X_{L} and X_{C} are the reactances associated with an L element and a C element, respectively. Consequently, the circuit consists of a series C_{L} and a shunt L_{L} element only.

This circuit demonstrates an antiparallel phase and group velocity, which is also known as left handedness, i.e., the E-field, H-field and Poynting vector make a left-handed triad. On the other hand, at high frequencies, C_L acts as an SC and L_L acts as an OC. As a result, the circuit consists of a series L_R and a shunt C_R element only, which is simply the model of a lossless TL. This circuit demonstrates a parallel phase and group velocity, which is also known as right handedness, i.e., the E-field, H-field and Poynting vector make a right-handed triad. Based on this logic, an MEBG is regarded as a composite right-/left handed (CRLH) structure. A CRLH structure is a special category of metamaterial, which has been used to design compact resonant antennas over the past two decades [LLI06].

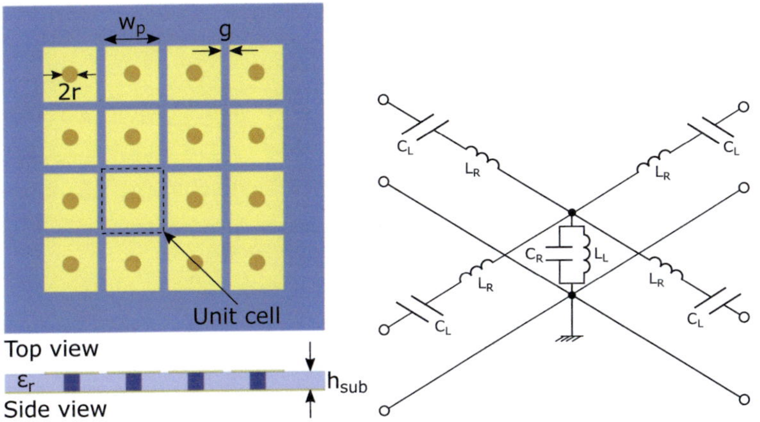

(a) 4×4 matrix of MEBG unit cells (b) Lumped element model of an MEBG unit cell

Figure 4.18: An MEBG periodic structure [CI06]

An MEBG is attractive for designing an antenna due to two reasons.

1 The dispersion curve, i.e., variation of the propagation constant with respect to the frequency, of an MEBG is neither unique nor linear. Therefore, its resonance frequencies do not exist in a harmonic ratio, as in the

case of a typical patch antenna. The design parameters of an MEBG can be optimized so as to produce a strong compression of its resonances in the desired frequency range [CI06]. This statement is described in detail in the following section.

2 An MEBG structure exhibits a series resonance ($\omega_{\mathrm{SE}} = 1/\sqrt{L_{\mathrm{R}}C_{\mathrm{L}}}$) and a shunt resonance ($\omega_{\mathrm{SH}} = 1/\sqrt{L_{\mathrm{L}}C_{\mathrm{R}}}$). In the frequency bandgap demarcated by ω_{SE} and ω_{SH} (provided $\omega_{\mathrm{SE}} \neq \omega_{\mathrm{SH}}$), the MEBG acts as a high impedance surface. Consequently, the surface waves in this frequency band are not bound to the surface, instead they are radiated as space waves [SZB+99], thus enhancing the radiation efficiency

4.2.2 Dispersion and Propagation Characteristics of an MEBG Unit Cell

For an EM wave propagating along the +x direction of an infinite one dimensional periodic structure, the E-field distribution ($\overrightarrow{E}$) at an arbitrary point x and at a later point $(x+np_{\mathrm{x}})$, where p_{x} denotes the structural periodicity along the x-axis and n is any positive or negative integer, are related to each other as per Eq. 4.4.

$$\overrightarrow{E}(x+np_x) = e^{-\gamma_x(\omega)np_x}\overrightarrow{E}(x) \tag{4.4}$$

In the above equation, $\gamma_{\mathrm{x}}(\omega)$ denotes the complex propagation constant. Further, $\gamma_{\mathrm{x}}(\omega) = \alpha_{\mathrm{x}}(\omega) + j\beta_{\mathrm{x}}(\omega)$, where $\alpha_{\mathrm{x}}(\omega)$ and $\beta_{\mathrm{x}}(\omega)$ denote the attenuation and phase constant along the x-axis, respectively. It is observed from this equation that if the periodic structure is lossless, i.e., $\alpha_{\mathrm{x}}(\omega) = 0$, then $\overrightarrow{E}$ at the points x and $(x+np_{\mathrm{x}})$ differ only in terms of a phase shift $e^{-j\beta_{\mathrm{x}}(\omega)np_{\mathrm{x}}}$. Therefore, $\beta_{\mathrm{x}}(\omega)$ is important for describing the EM wave propagation in a periodic structure. The dispersion curve for a plane wave in free space, $\beta(\omega)$ is given by Eq. 4.5.

$$\beta(\omega) = k(\omega) = \frac{\omega}{c_0} \tag{4.5}$$

In the above equation, k denotes the wavenumber and the $\beta(\omega)$ plot is simply a straight line, which is also known as the light line. In contrast, the dispersion characteristic plot of a one-dimensional periodic structure, $\beta_x(\omega)$, is obatined by solving an eigenvalue problem as shown in Eq. 4.6.

$$\vec{E}(x,y,z) = \sum_{n=-\infty}^{\infty} \vec{E_n}(y,z)e^{-j\beta_{\mathrm{xn}}(\omega)x} \quad , \quad \beta_{\mathrm{xn}}(\omega) = \beta_{\mathrm{x}}(\omega) + \frac{2n\pi}{p_{\mathrm{x}}} \tag{4.6}$$

The above equation is a simplified representation of the Bloch-Floquet's theorem, which states that an EM wave propagating in a periodic structure can be expressed as the sum of an infinite number of space harmonics. Since, the dispersion curve is repeated after an interval of $\frac{2\pi}{p_{\mathrm{x}}}$, it is sufficient to plot the dispersion curve within a single period, i.e., $0 \leq \beta_{\mathrm{x}} \leq \frac{2\pi}{p_{\mathrm{x}}}$, which is known as the Brillouin zone. This concept is extended to a two-dimensional MEBG structure. In a two-dimensional EBG, the EM wave propagates along both x- and y-axis. Therefore, Eq. 4.6 is extended to the y-axis as well by introducing a phase constant along the y-axis, i.e., $\beta_{\mathrm{y}}(\omega)$. It is important to note that in addition to the two-dimensional periodicity of an MEBG matrix, the geometry of an MEBG unit cell is also symmetric along the x- and y-axis. Therefore, the complete dispersion diagram of an MEBG unit cell is obtained by varying β_{x} and β_{y} along the triangular segment demarcated by the points Γ, X and M, as shown in Fig. 4.19a [23]. The triangular segment $\Gamma \rightarrow X \rightarrow M \rightarrow \Gamma$ is known as the irreducible Brillouin zone and the values of β_{x} and β_{y} at the points Γ, X and M are shown in Table 4.6. Using the CST eigenmode solver, the values of β_{x} and β_{y} are simultaneously varied along the irreducible Brillouin zone and the corresponding frequency eigenvalues are obtained, thereby plotting the dispersion diagram. The simulation model of the MEBG unit cell is shown in Fig. 4.19a [23]. The design parameter values are initially calculated using the guidelines given in [YS09], which uses $w_p = 0.1\lambda_0$, $g = 0.02\lambda_0$, $h_{sub} = 0.04\lambda_0$, $r = 0.005\lambda_0$ and $\varepsilon_r = 2.94$ for designing an MEBG at 4 GHz. The values calculated at 122 GHz are $w_p = 245$ µm, $g = 49$ µm, $h_{sub} = 98$ µm and $r = 12$ µm. The initial parameter values are optimized, since $h_{sub} = 105$ µm and $\varepsilon_r = 7$ are fixed for the selected LTCC substrate (DuPont 9K7). The optimized design parameter values used in the simulation are given in Table 4.7.

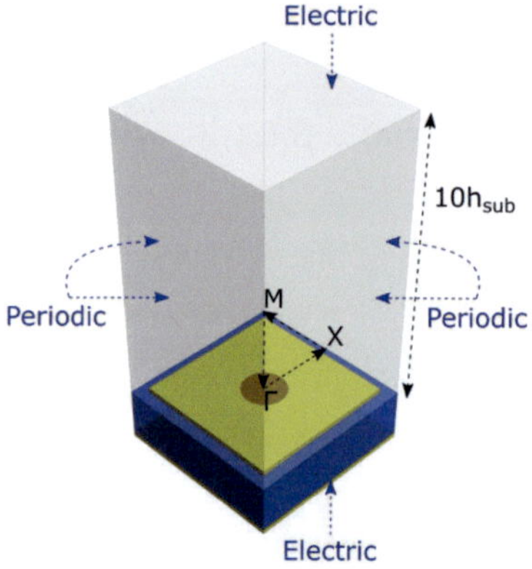

(a) Simulation model and boundary conditions

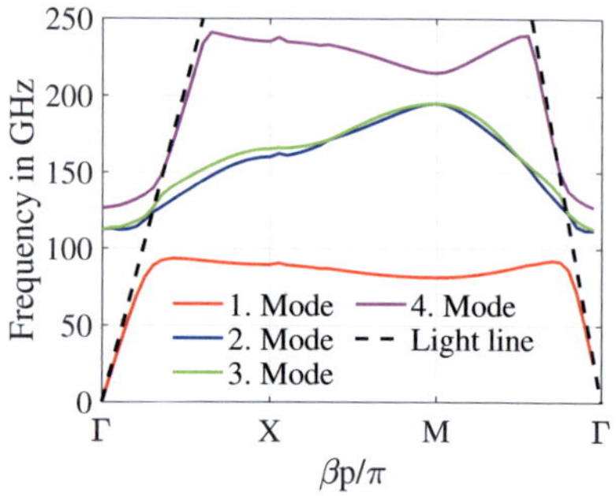

(b) Simulated dispersion diagram plot

Figure 4.19: MEBG dispersion diagram [23]

	β_x	β_y
Γ	0	0
X	π/p_x	0
M	π/p_x	π/p_y

Table 4.6: Irreducible Brillouin zone boundary values of an MEBG unit cell [23]

Further, the simulation employs periodic boundary condition on the four side walls of the unit cell, in order to emulate an infinite periodic structure. The top and bottom surface of the unit cell are terminated with an electric boundary condition (i.e., $\vec{E} = 0$), since open boundaries are not supported by the CST

Parameter	Value (µm)
w_p	335
g	52
r	50
h_{sub}	105

Table 4.7: MEBG design parameter values [23]

eigenmode solver. An air column of height $10h_{sub}$ is placed on the top of the MEBG unit cell so as to emulate a free space scenario [KU12]. The simulated dispersion diagram is shown in Fig. 4.19b [23]. It is seen that for a specific set of phase constants (β_x, β_y), distinct frequency eigenvalues exist which are known as modes. Each mode has a specific phase velocity, group velocity, E- and H-field distribution. The first four dispersion curves along with the light line are shown in the dispersion diagram (see Fig. 4.19b). In this diagram, the first two dispersion curves shown by the 1. Mode and 2. Mode show the TM wave and the TE wave, respectively [Sie99]. The dispersion curves shown by the 3. Mode and 4. Mode are higher-order right-handed modes [CI06]. The following conclusions are drawn using the dispersion diagram.

1 **Surface wave bandgap**: The dispersion curves of the surface wave modes lying below the light line are observed in order to determine the surface wave bandgap. The TM wave (shown by the 1. Mode in the dispersion diagram) shows an upper cutoff of 93 GHz and the TE wave (shown by the 2. Mode in the dispersion diagram) shows a lower cut-off of 123.5 GHz. The frequency bandgap between these two limits is known as the surface wave bandgap. The MEBG prevents propagation of any surface wave mode in this frequency range.

2 **Leaky wave modes**: In the dispersion diagram (see Fig. 4.19b), the dispersion curve segments, which lie below the light line (i.e., $\beta > k_0$,

where k_0 is the free space wavenumber) indicate waves, whose phase velocity is less than the speed of light (c_0). These waves do not radiate because their perpendicular propagation constant (β_y), given by Eq. 4.7, becomes imaginary. These waves are guided along the surface. In contrast, the dispersion curve segments, which lie above the light line (i.e., $\beta < k_0$) indicate waves, whose phase velocity is greater than the speed of light. In this case, the value of β_y is real. These waves are capable of radiation and they are termed as leaky waves [CI06] [Sie99] [SSLL02].

$$\beta_y = \sqrt{k_0^2 - \beta^2} \tag{4.7}$$

The angle of radiation of a leaky wave (ϕ) is given by Eq. 4.8 [CI06]. From this equation, it is observed that for a high frequency and for a small value of β, ϕ is negligible and hence the leaky wave radiates along the boresight direction.

$$\phi = \arcsin\left(\frac{\beta c_0}{2\pi f}\right) \tag{4.8}$$

The leaky wave resonance frequencies are determined by sampling the dispersion curve at intervals, shown by Eq. 4.9 [LLI06] [CI06]. In this equation, p and N denote the MEBG periodicity and the number of unit cells along the x- or y-axis, respectively.

$$\frac{\beta p}{\pi} = \pm\frac{n}{N} \quad , \quad n = 0, \pm 1, \pm 2 .... \pm (N-1) \tag{4.9}$$

4.2.3 MEBG Antennas with an MS-to-GCPW Feed

The MEBG structure, described in the previous section, is used to design two variants of an aperture-coupled MEBG antenna with an MS-to-GCPW feed. The first antenna variant employs a 4×4 matrix of MEBG unit cells on ML 1, as shown in Fig. 4.20. A ground plane aperture on ML 2 is placed below the center of the MEBG matrix. An MS and 50 Ω GCPW constituting the feedline are placed on the backside of the bottom substrate layer, i.e., on ML 3. The second antenna variant employs a 6×6 matrix of MEBG unit cells on ML 1. The structural design of its feed on ML 2 and ML 3 is the same as for the 4×4 MEBG antenna. This section aims at investigating the influence of the MEBG matrix size on the impedance bandwidth, realized gain and radiation efficiency of the antenna. The design parameters of the MEBG unit cell employed in both antenna variants are the same as shown in Table 4.7. The remaining design parameters for both antenna variants are given in Table 4.8. It should be noted that only the values of w_{ap} and l_{stub} are different in both variants.

Surface Wave Suppression with MEBGs

The MEBG unit cell is optimized to prohibit the propagation of surface waves in the frequency range of 93 to 123.5 GHz. The absolute E-field magnitude simulated on the top surface of the 4×4 MEBG and 6×6 MEBG antenna variants are shown in Fig. 4.21a and Fig. 4.21b, respectively. It is observed from these E-field plots that the suppression of surface waves achieved with the 4×4 MEBG and 6×6 MEBG antennas are nearly the same, which implies that the surface wave suppression depends only on the design of an MEBG unit cell and not on the overall size of its matrix. Further, on comparing these plots with the E-field plot of the aperture-coupled VF, SP antenna shown in Fig. 4.4a, it is seen that the efficiency of surface wave suppression achieved with the MEBG unit cells is similar to that of a VF structure. Moreover, the MEBG structure employs a thinner substrate and occupies a smaller area than the VF structure.

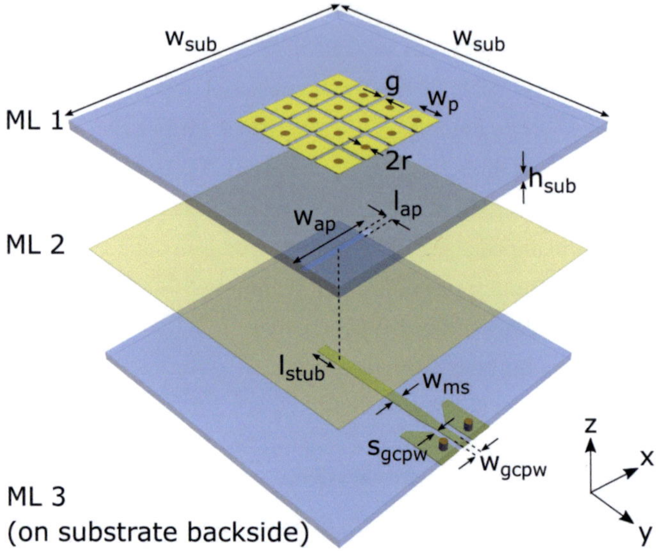

Figure 4.20: Multilayered structure of 4×4 MEBG antenna with MS-to-GCPW feed

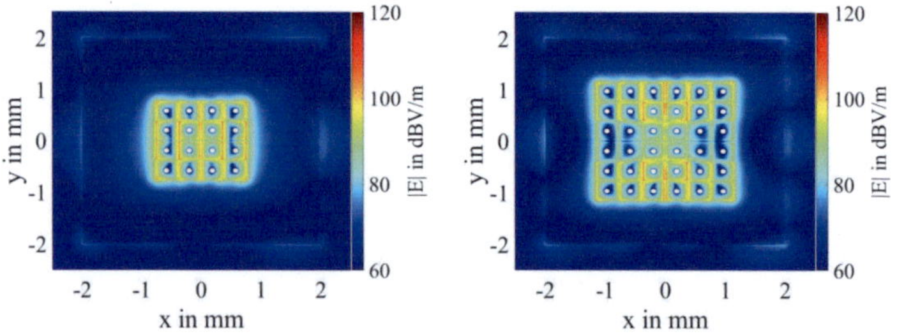

(a) 4×4 MEBG antenna with an MS-to-GCPW feed

(b) 6×6 MEBG antenna with an MS-to-GCPW feed

Figure 4.21: Simulated E-field of 4×4 and 6×6 MEBG antenna with MS-to-GCPW feed at 122 GHz

Parameter (see Fig. 4.20)	Value for 4×4 MEBG (µm)	Value for 6×6 MEBG (µm)
w_{sub}	4000	4000
l_{ap}	100	100
w_{ap}	1130	1550
l_{stub}	300	200
w_{ms}	147	147
w_{gcpw}	100	100
s_{gcpw}	66	66

Table 4.8: Design parameter values of 4×4 and 6×6 MEBG antenna with MS-to-GCPW feed

Estimation of Leaky Wave Resonance Frequencies in an MEBG

A detailed view of the first segement of the dispersion diagram, i.e., $\Gamma \rightarrow X$ segment of the irreducible Brillouin zone, is shown in Fig. 4.22a [23]. The leaky wave resonance frequencies of the 4×4 MEBG and 6×6 MEBG antennas are determined as follows. For the 4×4 MEBG antenna, the number of unit cells along the x- or y-axis is $N = 4$. On substituting this value in Eq. 4.9, we obtain $\frac{\beta p}{\pi} = 0$, 0.25, 0.50 and 0.75. Barring the 1. Mode, the eigenfrequencies on the remaining three dispersion curves, namely, 2. Mode, 3. Mode and 4. Mode are observed corresponding to the values $\frac{\beta p}{\pi} = 0$ and 0.25. As mentioned in section 4.2.2, the waves corresponding to these eigenfrequencies lie above the light line and hence they radiate as leaky waves. The eigenfrequencies for $\frac{\beta p}{\pi} = 0$ are observed at 112 GHz and 126 GHz. The eigenfrequencies for $\frac{\beta p}{\pi} = 0.25$ are observed at 118.2 GHz, 120.3 GHz and 134.9 GHz. Similarly, for the 6×6 MEBG antenna, $N = 6$ and the leaky wave resonance frequencies are the eigenfrequencies observed on the dispersion curves, corresponding to the values $\beta p/\pi = 0$ and 0.167. In this case, the leaky wave resonance frequencies are observed as 112.2 GHz, 115.2 GHz and 129.5 GHz.

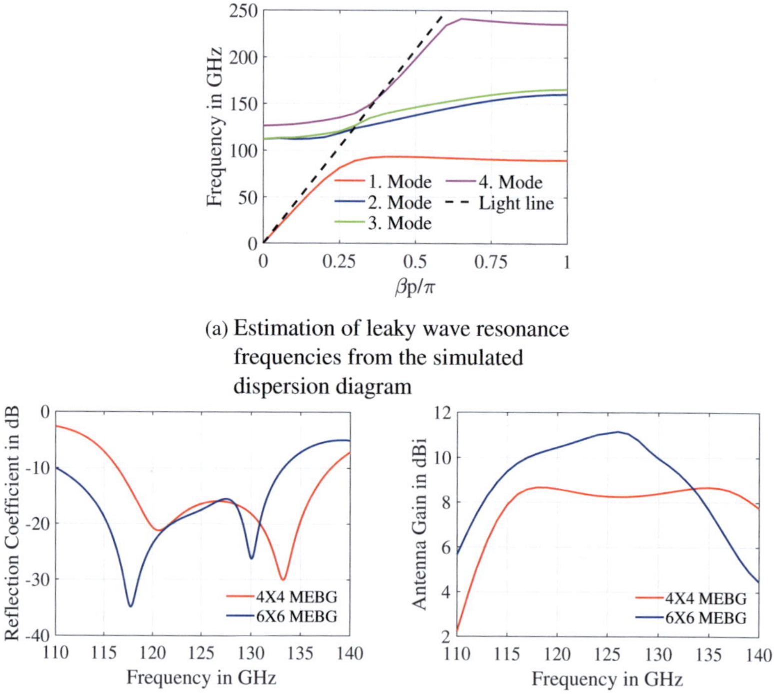

(a) Estimation of leaky wave resonance frequencies from the simulated dispersion diagram

(b) Simulated RC

(c) Simulated antenna gain

Figure 4.22: 4×4 and 6×6 MEBG antenna with MS-to-GCPW feed

Simulated Antenna Bandwidth and Realized Gain

Fig. 4.22b shows the simulated RCs of the 4×4 and 6×6 MEBG antennas. The simulated RC of the 4×4 MEBG antenna is less than -10 dB between 116.4 and 138.1 GHz, i.e., 17.05% relative bandwidth. The simulated RC of the 6×6 MEBG antenna is less than -10 dB between 110 and 133.3 GHz, i.e., 19.15% relative bandwidth. The sharp resonance peaks observed in the simulated RCs of the 4×4 and 6×6 MEBG antennas are compared with the resonance frequencies calculated using the dispersion diagram, as shown in Table 4.9.

Antenna variant	S-Parameter	Dispersion diagram
4×4 MEBG	120.5 GHz 133.3 GHz	118.2 GHz 120.3 GHz 134.9 GHz
6×6 MEBG	118 GHz 130 GHz	112.2 GHz 115.2 GHz 129.5 GHz

Table 4.9: Resonance frequencies observed in simulated RC curves versus dispersion diagram for 4×4 and 6×6 MEBG antenna with MS-to-GCPW feed [23]

A good agreement is observed between the two approaches. Note that the first two resonances observed for the 4×4 MEBG antenna in the dispersion diagram, i.e., 118.2 GHz and 120.3 GHz, lie close to each other and therefore it is difficult to distinguish them in the simulated RC curve. The same logic applies to the first two resonances of the 6×6 MEBG antenna as well. The simulated antenna gain of the 4×4 and 6×6 MEBG antennas are shown in Fig. 4.22c. The 4×4 MEBG antenna gain shows an antenna gain greater than 8.25 dBi within the −10 dB RC bandwidth, i.e., from 116.4 to 138.1 GHz. The antenna gain remains nearly constant in this frequency range and the maximum gain is 8.7 dBi. The simulated antenna efficiency is approximately 90% between 120 and 135 GHz and greater than 80% in the remaining frequency range. On the other hand, the antenna gain of the 6×6 MEBG antenna varies over its −10 dB RC bandwidth. The antenna gain is greater than 10 dBi between 117 and 130 GHz and the maximum gain is observed as 11.6 dBi at 126 GHz. The simulated antenna efficiency is greater than 85% between 113 and 125 GHz and greater than 80% in the remaining frequency range. The area occupied by the

4×4 MEBG and 6×6 MEBG radiating elements are 1.496 mm × 1.496 mm and 2.27 mm × 2.27 mm, respectively, whereas the size of the radiating element in the VF, SP antenna, shown in section 4.1.1, is 3.1 mm × 3.1 mm. Therefore, the 4×4 MEBG and 6×6 MEBG antennas show a size reduction of 51.74% and 26.77%, respectively. Additionally, the antenna substrate of the 4×4 and 6×6 MEBG antennas is only half as thick as the antenna substrate of the VF antennas. In spite of its thin and compact dimensions, the 4×4 MEBG antenna shows an antenna gain of 8.4 dBi at 122 GHz, which is comparable to the VF, SP antenna gain. The 6×6 MEBG antenna shows an antenna gain of 10.7 dBi at 122 GHz, which is higher than the VF, SP antenna gain. Based on this investigation, it is concluded that if the size of the MEBG matrix is increased beyond 6×6 MEBG, the frequency range over which a higher antenna gain is achieved would further reduce. Since the MMIC selected in this work operates between 119.3 and 125.8 GHz, a 6×6 MEBG antenna offers an excellent compromise between the impedance bandwidth and the realized antenna gain.

4.2.4 MEBG Antennas with an SL-to-GCPW Feed

Two MEBG antenna variants with an SL-to-GCPW feed using 4×4 and 6×6 MEBG periodic structures are designed. The multilayered structure of the 6×6 MEBG antenna with an SL-to-GCPW feed is shown in Fig. 4.23 [23]. The design of the 4×4 MEBG antenna variant differs only in terms of its radiating element on ML 1. Each antenna variant consists of four layers of DuPont 9K7PV substrate and five layers of Au metal. These antenna variants were initially designed using three substrate layers and four metal layers (i.e., one substrate layer for the antenna substrate and two substrate layers for the feed network), but the antennas are intentionally made thicker in order to avoid the risk of warpage during the LTCC manufacturing. As shown previously in section 4.1.1, the antenna had three substrate layers and therefore its ground plane was replaced by a coarse metal mesh, in order to avoid warpage during

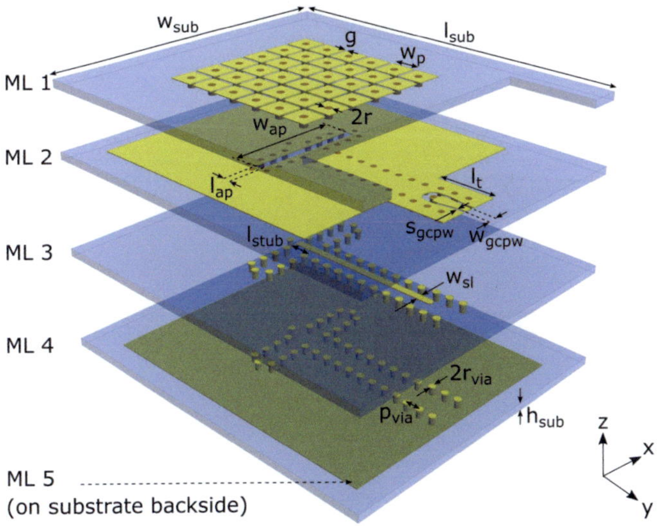

Figure 4.23: Multilayered structure of 6×6 MEBG antenna with SL-to-GCPW feed [23]

the LTCC co-firing process. The radiating elements on ML 1 of the antenna variants are the same as in the previous section on 4×4 and 6×6 MEBG antennas with an MS-to-GCPW feed. The feed network on ML 2, ML 3 and ML 4 is similar to the feed network shown in section 4.1.2. Further, ground vias are used around the aperture (on ML 2) and on either side of the SL signal conductor (on ML 3) in order to suppress an undesired radiation and coupling. The optimized values of design parameters for both antenna variants are shown in Table 4.10. The principle of an MEBG matrix and an SL-to GCPW feed have been already discussed in previous sections. The manufactured prototypes of the 4×4 MEBG and 6×6 MEBG antenna with an SL-to-GCPW feed are shown in Fig. 4.24a and Fig. 4.24b, respectively [23]. The antennas are measured using a 150 μm pitch GSG probe. The measurements are preceeded by a gain calibration and an SOL calibration (same as in section 4.1). The measurement and simulation results of the antennas are discussed next.

Parameter (see Fig. 4.23)	Value in µm (4×4 MEBG)	Value in µm (6×6 MEBG)
w_{sub}	3800	4300
l_{sub}	4600	5100
l_{ap}	90	120
w_{ap}	940	1580
w_{gcpw}	100	100
s_{gcpw}	41	41
l_t	800	800
l_{stub}	220	150
w_{sl}	100	100
r_{via}	80	80
p_{via}	200	200
h_{sub}	105	105

Table 4.10: Design parameters of 4×4 and 6×6 MEBG antenna with SL-to-GCPW feed [23]

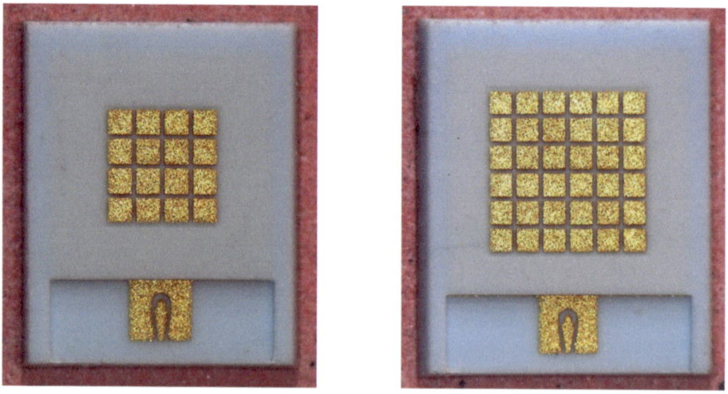

(a) 4×4 MEBG antenna

(b) 6×6 MEBG antenna

Figure 4.24: Prototypes of MEBG antenna with SL-to-GCPW feed [23]

4X4 MEBG antenna with an SL-to-GCPW feed: The measured and simulated RCs are shown in Fig. 4.25a [23]. In both measurement and simulation, it is observed that the resonance peak of the antenna lies at 122 GHz. The measured RC is less than −10 dB between 119 and 124 GHz, i.e., 4.11% relative bandwidth. The measured RC at 122 GHz is −13.5 dB. Although the progression of the measured and simulated RC curves is similar, the measured impedance matching shows a significant amount of degradation in comparison to the simulation. The measured and simulated realized antenna gain over frequency curves are shown in Fig. 4.25b [23]. Over the −10 dB RC bandwidth, the measured antenna gain varies between 8 dBi and 9.86 dBi. The simulated antenna efficiency in this frequency range is close to 85%. The measured and simulated radiation pattern at 122 GHz in H- and E-plane are shown in Fig. 4.25c and Fig. 4.25d, respectively [23]. The measured antenna gain along the boresight direction is 9.3 dBi, which is only 0.3 dB less than the simulated value. In both H- and E-plane, a good agreement is observed between the measured and simulated radiation patterns.

6X6 MEBG antenna with an SL-to-GCPW feed: The measured and simulated RCs are shown in Fig. 4.26a [23]. In the measured RC, the first resonance peak occurs at 116 GHz, which matches very well with the simulated RC. In addition, the second resonance peak occurs at 125.6 GHz in the measured RC, whereas in the simulated RC, it is observed at 129 GHz. The RC measured at these two resonance frequencies is extremely good, i.e., −22 dB at 116 GHz and −24 dB at 125.6 GHz, but the RC measured at around 122 GHz is significantly less than in the simulated RC. The measured and simulated realized antenna gain versus frequency curves are shown in Fig. 4.26b [23]. The measured antenna gain is greater than 10 dBi between 115.5 and 124.5 GHz, i.e., 7.5% relative bandwidth. The simulated antenna efficiency in this frequency range is greater than 80%. Therefore, a high antenna gain is achieved over the entire frequency range of interest. The measured peak antenna gain is 12.3 dBi, at 119.3 GHz. The measured antenna gain at 122 GHz is 10.4 dBi.

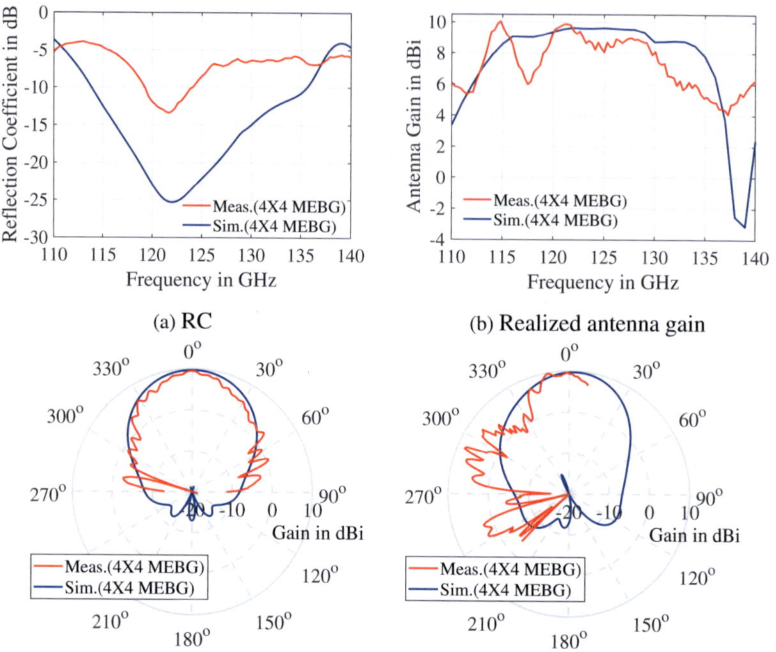

(a) RC

(b) Realized antenna gain

(c) Farfield radiation pattern in H-plane at 122 GHz

(d) Farfield radiation pattern in E-plane at 122 GHz

Figure 4.25: Measurement and simulation of 4×4 MEBG antenna with SL-to-GCPW feed [23]

On comparing the measured and simulated gain curves, it is seen that the measured gain curve shows a sharp degradation beyond 125 GHz. This sudden degradation in the antenna gain is also visible in the simulation, although it occurs at around 130 GHz. The measured and simulated radiation patterns of the antenna at 122 GHz in H- and E-planes are shown in Fig. 4.26c and Fig. 4.26d, respectively [23]. The measured antenna gain along the boresight direction is 10.45 dBi, which is approximately 1.5 dB less than in the simulation. Barring the probe influence, a good agreement is achieved between the shape of the measured and simulated radiation patterns in both H- and E-plane.

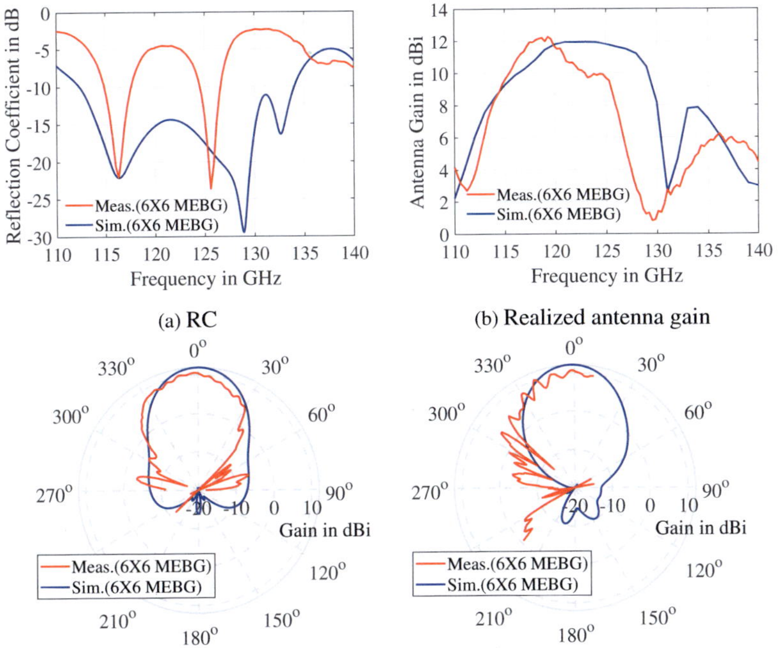

(a) RC

(b) Realized antenna gain

(c) Farfield radiation pattern in H-plane at 122 GHz

(d) Farfield radiation pattern in E-plane at 122 GHz

Figure 4.26: Measurement and simulation of 6×6 MEBG antenna with SL-to-GCPW feed [23]

Influence of Layer Misalignment

A prototype of the aperture-coupled 6×6 MEBG antenna with an SL-to-GCPW feed is analyzed using X-ray. Fig. 4.27a shows the X-ray image [23]. The following two observations are made from this image. First, the MEBG vias do not lie exactly at the center of the MEBG metal patches. Second, the ground plane aperture does not lie exactly below the center of the 6×6 MEBG matrix. In fact, the ground plane aperture appears to be shifted along the y-axis. Next, the prototype is analyzed using X-ray microtomography. Two resultant images are shown in Fig. 4.27b and Fig. 4.27c [23]. Fig. 4.27b shows the 6×6 MEBG antenna along the middle YZ plane. In this image, besides the asymmetric SL signal conductor on ML 3 and the ground plane aperture on ML 2, the MEBG patches on ML 1 are partly visible. The presence of MEBG patches in this image indicates an anomaly because as per the simulation model (see Fig. 4.22), the asymmetric SL signal conductor and the ground plane aperture should lie below the central vertical slot of the MEBG matrix. Fig. 4.27c shows the 6×6 MEBG antenna along a YZ plane, which passes through a vertical column of the MEBG unit cells. It is observed from this image that the midpoints of the ground plane aperture and the central slot of the MEBG structure are not properly aligned along the y-axis, thus indicating a layer misalignment along the y-axis.

The influence of layer misalignment on the RC and realized gain of the 6×6 MEBG antenna is investigated in simulation. In this simulation study, the layers ML 1, ML 2, ML 3 and ML 4 are misaligned one after the other, in steps of 20 µm, 40 µm and 60 µm, in both H-plane (XZ-plane) and E-plane (YZ-plane). It is observed that a significant deviation in the antenna performance occurs when either the ML 2 or the ML 3 are misaligned along the E-plane, i.e., the YZ-plane. The simulation result of these two cases are shown in Fig. 4.28 [23]. In Fig. 4.28a, it is seen that as the misalignment of ML 2 increases from 20 to 60 µm, the first resonance remains nearly fixed at around 116 GHz, whereas the second resonance shifts from 128 to 125.5 GHz. Additionally, the RC at 122 GHz degrades from −15 to −8 dB. A strong similarity in terms of resonance frequencies and degradation in the impedance matching at 122 GHz is observed between the simulated RC curve for $\Delta y_2 = 60$ µm and the measured RC of the 6×6 MEBG antenna, shown in Fig. 4.25a. In terms of the realized antenna gain (see Fig. 4.28b), as the misalignment increases from 20 to 60 µm, the bandwidth over which the antenna achieves a realized gain greater than 10 dBi decreases. For $\Delta y_2 = 20$ µm, the antenna gain is greater than 10 dBi between 116 and 129 GHz, whereas for $\Delta y_2 = 60$ µm, the antenna gain is greater than 10 dBi between 117 and 125 GHz. Fig. 4.28c shows the influence of ML 3 misalignment on the simulated RC of the 6×6 MEBG

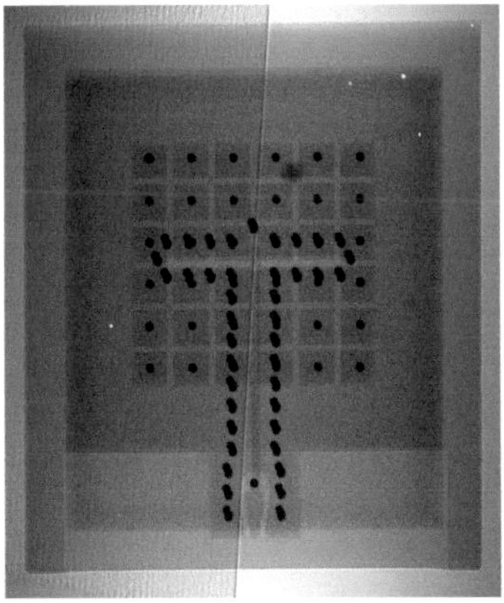

(a) Top view

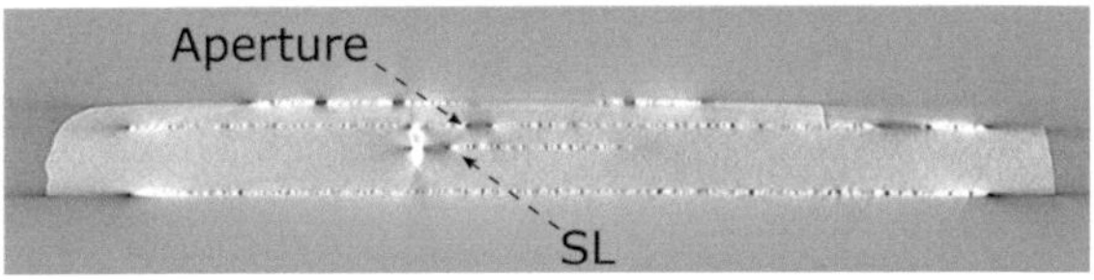

(b) Side view of the middle YZ plane

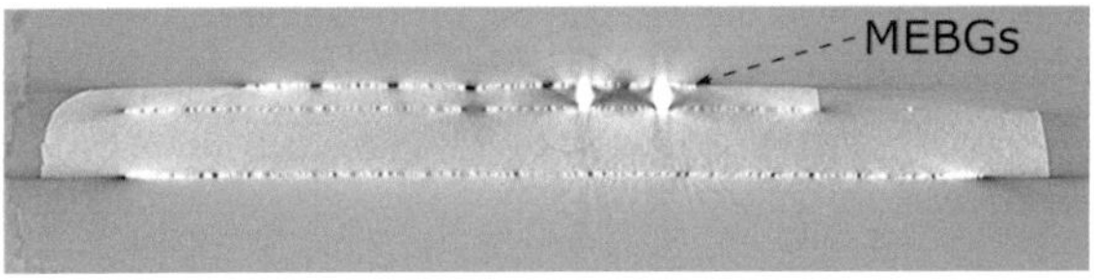

(c) Side view of a YZ plane crossing an MEBG column

Figure 4.27: X-ray images of a prototype of 6×6 MEBG antenna with SL-to-GCPW feed [23]

antenna. As the ML 3 misalignment increases from 20 to 60 µm, the simulated RC at 125 GHz degrades from -14.5 to -5 dB and both resonances are shifted to lower frequencies. In terms of realized antenna gain (see Fig. 4.28d), for $\Delta y_3 = 20\,\mu\text{m}$, the antenna gain is greater than 10 dBi between 116 and 129 GHz and a sharp dip in gain is observed at 131 GHz. For $\Delta y_3 = 60\,\mu\text{m}$, the antenna gain is greater than 10 dBi between 116 and 124 GHz and a sharp dip in gain is osberved at 126 GHz. A strong similarity is observed in the gain versus frequency curve correponding to the parameter value ($\Delta y_3 = 60\,\mu\text{m}$) and the measured gain versus frequency curve of the 6×6 MEBG antenna shown in Fig. 4.26b. On the basis of this simulative study, it is concluded that the deviation observed in the measurement and simulation result of the 6×6 MEBG antenna prototype has principally occurred due to the misalignment of the asymmetric SL feed line and the ground plane aperture with respect to the MEBG structure along the y-axis.

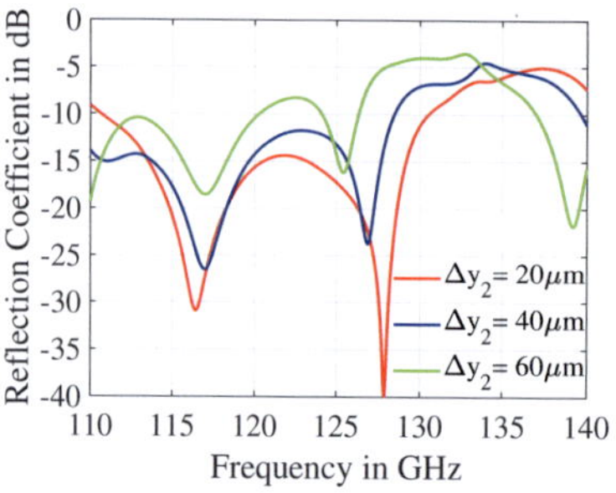

(a) Influence of ML 2 misalignment on the simulated RC

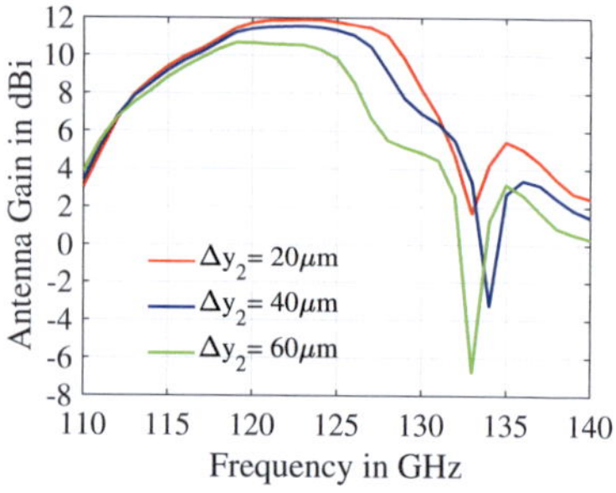

(b) Influence of ML 2 misalignment on the simulated antenna gain

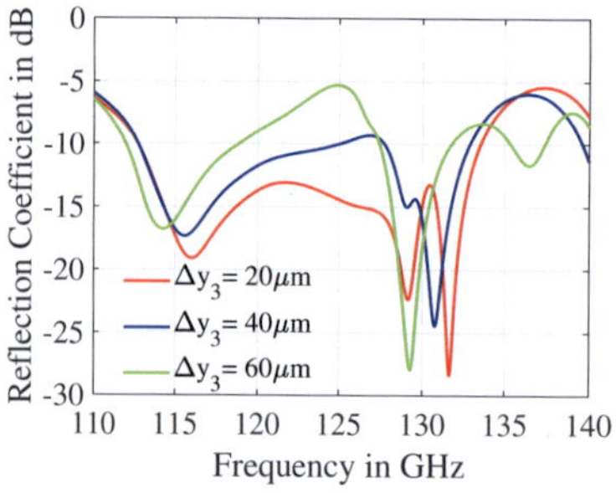

(c) Influence of ML 3 misalignment on the simulated RC

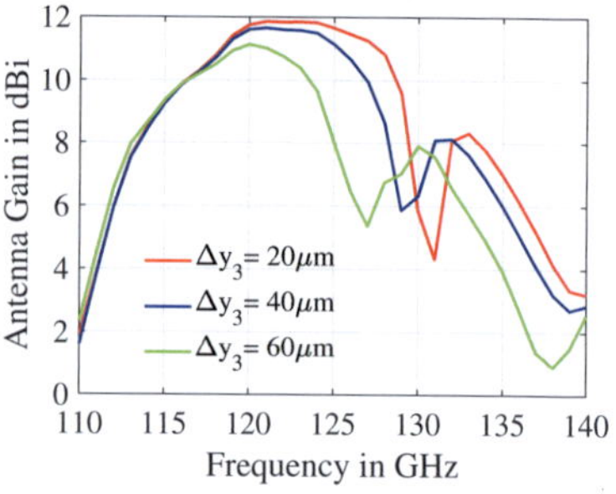

(d) Influence of ML 3 misalignment on the simulated antenna gain

Figure 4.28: Simulated influence of layer misalignment for 6×6 MEBG antenna with SL-to-GCPW feed [23]

4.3 Summary

In this chapter, two different aperture-coupled antenna concepts, namely a VF-based and an MEBG-based antenna, have been implemented in varied configurations. The antenna substrate of the VF antenna variants is made up of two DuPont 9K7PV layers, which leads to a cutoff frequency of 145 GHz for the TE_1 surface wave mode. Therefore, only the TM_0 surface wave mode propagates in the target frequency range. The surface wave power loss can be neglected, provided the antenna substrate thickness is less than 44 μm at 122 GHz. Since this is not the case here, it becomes essential to suppress the surface waves. The VF shown in this chapter is a modern realization of a soft surface, which is used to suppress the TM_0 surface wave mode. On the basis of simulation results, it is established that the VF achieves an adequate suppression of surface waves. In addition, it improves the farfield radiation pattern by increasing the boresight antenna gain, reducing the side lobe and back lobe radiation levels. Moreover, the VF acts as a resonator and it increases the impedance bandwidth of the antenna. Two VF antennas with different feed configurations are demonstrated, namely an aperture-coupled VF, SP antenna with an MS-to-GCPW feed and an SL-to-GCPW feed. The measurement results of both antennas are in good agreement with their corresponding simulation results. Subsequently, an aperture-coupled array with 2×2 VF patch antennas and an SL network-to-GCPW feed is presented. The array elements are separated by a distance of 1.55 mm (i.e., $0.63\lambda_0$ at 122 GHz). The array feed network consists of three quarter-wave matched SL power dividers, four 90° SL bends and a via-based SL-to-GCPW signal transition. The losses incurred by the array feed network are calculated by simulating the array with an ideal (lossless) feed network. A comparison of the ideal and real array gain shows that the feed network results in a loss of 1.8 dB at 122 GHz. Out of this value, a loss of 0.25 dB occurs due to the via-based SL-to-GCPW signal transition and the remaining loss of 1.55 dB occurs due to the SL feed network. The measurement and simulation results of the array are in good agreement with each other. As per the measurement result, the array achieves an RC less than −10 dB between 118 and 126.5 GHz and an antenna gain greater than 10 dBi between 119.3 and 129.5 GHz. The array gain measured at 122 GHz is 11.3 dBi and the maximum array gain is measured as 12.85 dBi at 124.4 GHz. The simulated efficiency of the antenna is greater than 70% between 117 and 130 GHz. To conclude, the array shows an adequate RC and antenna gain over the complete frequency range of the MMIC selected in this work (i.e., 119.3 to 125.8 GHz). Therefore, the array is suitable for realizing a horizontal-configuration LTCC SiP at 122 GHz.

Next, an MEBG-based antenna concept is presented. At first, an MEBG unit cell is analyzed on the basis of its lumped element model, which indicates that an MEBG unit cell is similar to a CRLH structure. Thereafter, the MEBG unit cell design parameters, namely, w_p, g, r and h_{sub} (see Fig. 4.18a) are optimized by simulating its dispersion diagram, $\beta(\omega)$ in the CST eigenmode solver. The optimization is done on the basis of the following two criteria. First, a surface wave bandgap should lie around 122 GHz. The MEBG unit cell acts as a high impedance surface in the bandgap, thereby preventing surface wave propagation and increasing the radiation efficiency. Second, a strong compression of the leaky wave resonances should be obtained in the desired frequency range. After optimizing the MEBG unit cell geometry, a 4×4 and 6×6 MEBG antenna using an MS-to-GCPW feed are investigated in simulation. Subsequently, a 4×4 and 6×6 MEBG antenna using an SL-to-GCPW feed are manufactured and their measurement and simulation results are compared. As per the measurement result, a 4×4 MEBG antenna with an SL-to-GCPW feed shows an RC less than −10 dB between 119 and 124 GHz. The antenna gain in this frequency range varies between 8.2 dBi and 9.5 dBi. At 122 GHz, the measured RC is −13.5 dB and the measured antenna gain is 9.3 dBi. In contrast, the measured antenna gain of a 6×6 MEBG antenna varies between 10 dBi and 12.3 dBi in the frequency range of 115.5 to 124.5 GHz. The measured antenna gain at 122 GHz is 10.44 dBi. The measured RC of the antenna shows a significant amount of deviation with respect to the simulation, which is investigated by means of X-ray imaging. The X-ray images and a simulative study on layer misalignment indicate that the SL feed line and the ground plane aperture are misaligned by approximately 60 µm along the E-plane, which leads to a degradation of the measured RC at around 122 GHz. In comparison with the 2×2 VF patch array, the 6×6 MEBG antenna substrate is only half as thick, yet the measured peak antenna gain of both antennas are comparable, i.e., 12.85 dBi for the 2×2 VF patch array and 12.3 dBi for the 6×6 MEBG antenna. Moreover, the bandwidth over which the 6×6 MEBG antenna achieves an antenna gain greater than 10 dBi is also comparable to the 2×2 VF patch array. On the downside, the 6×6 MEBG antenna design is found to be more susceptible to the layer misalignment, which usually occurs during the LTCC manufacturing process. In contrast, the 2×2 VF patch array design proves to be robust enough to withstand the typical LTCC tolerances.

5 LTCC SiP with Molding Encapsulation

A semiconductor MMIC consists of a large number of active and passive components densely integrated in a 3D structure with multiple layers of metallization separated by the semiconductor layers. The top metallization layer of an MMIC consists of several I/O pads, which are protected by a relatively thin passivation layer made up of silicon dioxide (SiO_2), silicon nitride (Si_3N_4) or the like. The passivation layer acts as the first line of defence, which protects the delicate semiconductor circuits against atmospheric degradation [TRK99]. Further, in a surface-mount SiP, such as the horizontal-configuration SiP concept shown in section 2.3.2, the I/O pads on the MMIC are connected to the off-chip components using fragile wirebonds or flip-chip bumps with a diameter in the range of tens of micrometers. Therefore, it is important to encapsulate the SiP, in order to protect the MMIC as well as its fragile I/O interconnects against atmospheric contaminants (e.g., moisture, ions and gases) and mechanical damage. The package encapsulation plays an important role in ensuring that an SiP delivers a stable and reliable performance under various environmental conditions over an extended period of time.

5.1 A Near-Hermetic Package

There are two types of electronic packages, which are commonly used today for consumer applications, namely, hermetic and near-hermetic packages. A hermetic package strictly restricts the moisture, ionic contaminants and other harmful gases from permeating the package. The permeation rate is defined by a universal standard named Mil-STD-883 Test Method 1014 [Ass16]. The hermetic encapsulation concept is usually implemented by using a lid made up of metal, ceramic or glass, as shown in Fig. 5.1a. If the lid-based concept is used to encapsulate the horizontal-configuration SiP (shown in section 2.3.2), then the lid should be made up of glass or ceramic and the lid thickness should be $\lambda_g/2$ (λ_g is the guided wavelength in the lid material at the frequency of interest). The latter condition ensures that the lid has no negative influence on the radiation characteristics of the antennas transmitting and receiving the signal of the MMIC transceiver [Bee13]. This approach entails a high cost and a huge manufacturing effort. Consequently, it is not well suited for a high-volume consumer application. In contrast, a near-hermetic package, although it is not fully-hermetic, it is good enough to meet the performance requirements and the expected lifetime in a typical end-user environment [Ass16]. A near-hermetic package utilizes organic materials, e.g., polymers and epoxy resins, for encapsulating a package. Two near-hermetic package concepts, namely, a glob-top and a cavity-filling package encapsulation are shown in Fig. 5.1b and Fig. 5.1c, respectively. In a glob-top package encapsulation, an adequate amount of a liquid molding compound is dispensed on the top of an SiP followed by curing of the molding compound. The molded device after curing has a dome shape. In this case, the liquid molding compound, depending on its viscosity, flows freely over the SiP and therefore the encapsulated volume cannot be strictly controlled [Inc14]. Next, a cavity-filled package encapsulation is implemented on an SiP, which has a surface cavity of a predefined size and shape. The encapsulation process is same as the glob-top encapsulation, but in this case, the cavity walls restrict the flow of the liquid molding compound, thus producing a uniform encapsulant profile [Tum01]. Moreover, in a cavity-filled package encapsulation, liquid molding compounds with a lower viscosity can be used, which helps in reducing the entrapment of voids during the encapsulation process. In addition, it is also possible to fabricate the package cavity by using a high-viscosity (i.e., non-flowing) molding compound. The high-viscosity molding compound is cured and subsequently, the package cavity is filled with a lower-viscosity molding compound. This concept is known as dam-and-fill encapsulation [Asy].

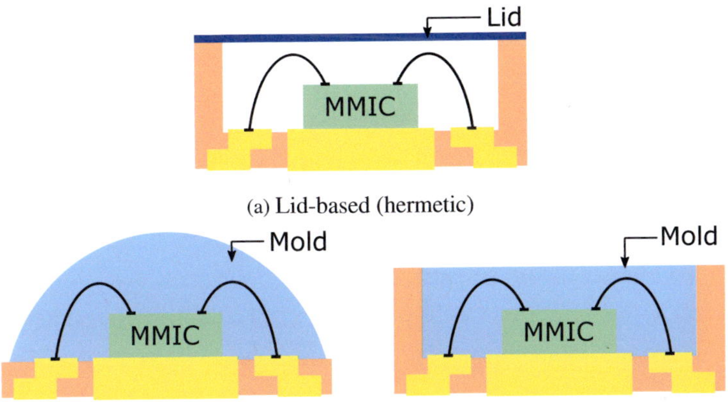

(a) Lid-based (hermetic)

(b) Glob-top (near-hermetic)

(c) Cavity-filling (near-hermetic)

Figure 5.1: Types of package encapsulation

In this work, a cavity-filled encapsulation concept is used due to the following reasons. First, molding compounds are usually cheaper than hermetic lids. Second, the encapsulation process is simple and it is commonly used for producing electronic packages on a mass scale. Third, in comparison with a glob-top encapsulation, a cavity-filled encapsulation results in a well-defined volume of the encapsulant, which helps in building a simulation model of the encapsulated SiP. Therefore, the influence of the molding encapsulation on the radiation characteristics of the antennas, e.g., RC, realized gain and farfield radiation pattern, are determined on the basis of EM simulations. Moreover, a thermal simulation of a horizontal-configuration LTCC SiP is performed in order to estimate the maximum operating temperature of the selected MMIC.

5.1.1 Selection of a Molding Encapsulant

The selection of a molding compound for encapsulating a mm-wave horizontal configuration SiP is based on the following criteria.

1 Thermomechanical Properties: A molding encapsulation leads to two types of residual stress. First, a mechanical stress, which develops due to the shrinkage of the molding compound during the curing process. This stress is reduced by adding suitable modifiers to the molding compound. Second, a thermal stress, which develops due to the mismatch of the CTEs of the molding compound and various SiP components. This mismatch occurs when the molding compound cools down from the curing temperature to the room temperature. The thermal stress is reduced by selecting a molding compound with a CTE as close as possible to the semiconductor material and the LTCC substrate, and a low value of elasticity modulus. A reduction in the CTE is achieved by adding fillers to the molding compound (e.g., glass), which in turn increases the elasticity modulus [BBR+90] [TRK99] [Tum01]. Therefore, a tradeoff is required between these two properties while selecting a molding compound [BBR+90].

2 Thermal properties: The molding compound should be thermally conductive, since it helps in dissipating the heat generated by the MMIC. The most commonly used molding compounds are epoxy resins. The thermal conductivity of an epoxy resin is increased by adding suitable filler material to it. The standard fillers used in a thermally conductive and electrically insulating epoxy resin are Al, Al_2O_3 and boron nitride (BN) [Gmb12]. It is intended to mount the encapsulated SiP on a PCB by using reflow soldering process, during which the temperature can go up to 235 °C [Com18]. Therefore, the maximum operating temperature of the molding compound should also be considered.

3 Physical Properties: There are two important physical properties that should be considered while selecting a molding compound, namely, viscosity and adhesion [Tum01]. If a molding compound has a low viscosity, then it flows easily around the SiP components, e.g., MMIC and wirebonds, without creating any voids. The viscosity of a molding

compound is not constant and it can be temporarily lowered by either heating the molding compound to a temperature of 40 to 50 °C or by using a thixotropic molding compound. A thixotropic molding compound shows a lower viscosity when a shear force is applied to it (e.g., when it is dispensed in the SiP cavity) and a higher viscosity under static condition [Gmb12]. The second physical property is adhesion. If the adhesion between the molding compound and the SiP components is good, it helps in maintaining thermal integrity of the SiP under thermal stress and the crack resistance of the molding compound is strengthened [BBR+90]. Additionally, the molding compounds, which do not require special storage conditions and are suitable for manual application, are given preference in this work.

4 Dielectric properties: According to the horizontal-configuration SiP concept, the via-based SL-to-GCPW signal transition and the mm-wave wirebond interconnects are encapsulated in the molding compound. Therefore, the dielectric properties (ε_r, $\tan\delta$) of the molding compound should be characterized in the desired frequency range and the $\tan\delta$ value should be as low as possible in this frequency range. In contrast, the dielectric properties of commercial molding compounds are usually available in the kHz or MHz range only.

Taking the above mentioned requirements into consideration, an epoxy resin, Polytec TC 430-T is selected for encapsulating the horizontal - configuration LTCC SiP. The main features of Polytec TC 430-T, which justify its suitability for the task at hand, are as follows. First, it has an excellent thermal stability and hence the ability to withstand thermal stress. Second, it has fine BN fillers, which lead to a good thermal conductivity. Third, its maximum operating temperature is higher than the maximum temperature required for the reflow soldering process, i.e., 225 to 235 °C [Com18]. Fourth, it is a soft thixotropic paste, which is prepared by mixing two components. Both mixing and dispensing processes can be done manually. Moreover, no special conditions are required for its storage. Finally, the dielectric properties of Polytec TC 430-T, provided by the manufacturer are specified at a very low frequency of 1 kHz [Gmb15]. Recently, the dielectric properties of Polytec TC 430-T have been measured in the frequency range of 110 to 170 GHz, by using a commercial material characterization kit from SWISSto12 [GWB+18]. The values measured at 122 GHz, $\varepsilon_r = 3.185$ and $\tan\delta = 0.023$, are used in all subsequent simulations involving Polytec TC 430-T. The most important properties of Polytec TC 430-T are summarized in Table 5.1 [Gmb12].

Thermomechanical Properties	
Youngs modulus	5.6 GPa
CTE (Uncured state)	26 ppm/K
CTE (Cured state)	135 ppm/K
Thermal Properties	
Thermal conductivity	0.7 W/mK
Maximum temperature	400 °C
Physical Properties	
Viscosity (Uncured state)	28 000 mPa.s
Adhesion strength	11 MPa
Dielectric Properties at 122 GHz	
ε_r	3.185
$\tan\delta$	0.023

Table 5.1: Properties of the selected molding encapsulant - Polytec TC 430-T

5.1.2 Influence of Molding Encapsulation: A Proof of Concept

The selected molding compound, Polytec TC 430-T, is used to encapsulate the following components of the LTCC-based horizontal-configuration SiPs shown in this work: via-based SL-to-GCPW signal transitions of the mm-wave module, MMIC, mm-wave and baseband wirebond interconnects. The influence of the molding compound on the RC, realized gain (versus frequency) and farfield radiation pattern of the antenna are determined by building a proof of concept. The proof of concept, in its most basic form, consists of an antenna under test (AUT), a dummy chip with a 50 Ω GCPW and three parallel half-wavelength wirebonds in GSG configuration, which interconnect the AUT and dummy chip. Two separate wirebonded assemblies of the AUT and dummy chip are manufactured. In the first assembly, the AUT-to-dummy chip mm-wave interconnect is encapsulated with a specific thickness of Polytec TC 430-T. In the second assembly, the AUT-to-dummy chip mm-wave interconnect is left unencapsulated. In both cases, the RC, realized gain (versus frequency) and farfield radiation pattern (at 122 GHz) of the antenna are simulated as well as measured using a GSG probe. The AUT used in both encapsulated and unencapsulated assemblies is a VF, SP antenna with an SL-to-GCPW feed, whose multilayered structure, simulation and measurement results are shown in section 4.1.2. The dummy chip used in both assemblies consists of a 50 Ω GCPW line on an Al_2O_3 substrate ($\varepsilon_r = 9.9$, $\tan\delta = 0.0001$) with a thickness of 254 µm. The total size of the dummy chip is 2 mm×2 mm and the length of the GCPW is 1.9 mm. The steps involved in building the encapsulated and unencapsulated AUT and dummy chip assembly are as follows.

1 The AUT and dummy chip are glued on a metal plate at an edge-to-edge distance of around 560 µm by using a flip-chip bonder. Next, a semi-automatic wire bonder is used to form three wedge-wedge Au wirebonds in GSG configuration between the AUT and dummy chip (see Fig. 5.2a) [22]. Each wirebond has a diameter of 17.5 µm and a maximum loop height of approximately 110 µm from the antenna's bond pad. The resulting wirebonds have a loop length of approximately $\lambda_g/2$ (i.e., 689 µm) at 122 GHz in Polytec TC 430-T as medium. On the

dummy chip side, the center-to-center distance between the signal conductor and the ground plane (on either side) is 100 μm and on the AUT side, the distance between the center of the signal pad and the ground pad edge (on either side) is 91 μm. Therefore, although it is desired, the three GSG wirebond interconnects are not exactly parallel to each other in this case.

2 A Rohacell ($\varepsilon_r \approx 1$) structure is placed on the wirebonded assembly of the AUT and dummy chip (see Fig. 5.2b) [22]. Hence, an air cavity is formed, which is bounded by the top two substrate layers of the AUT on one side and the Rohacell structure on the remaining three sides. The formation of this air cavity is essential because it ensures that the AUT and dummy chip assembly is encapsulated with a uniform height of the molding compound. Further, it restricts a free flow of the liquid molding compound. Hence, the GSG pads on the other end of the dummy chip are free for probe-based measurement.

3 A thixotropic paste of the molding compound, Polytec TC 430-T, is prepared by mixing its (two) components in a specific ratio, as described in the manufacturer's data sheet [Gmb15]. Thereafter, the paste is dispensed in the air cavity. The result of this step is shown in Fig. 5.2c [22]. Both mixing and dispensing of the molding compound are done manually. Therefore, voids may be formed due to entrapment of air during the molding process. Consequently, the ε_r of the molding compound may slightly vary depending on the volume percentage of air trapped in the molding encapsulation. Such errors can be minimized or avoided in an automated production scenario.

4 The molded AUT and dummy chip assembly is cured at a temperature of 150 °C for nearly of 15 min, as per the manufacturer's data sheet [Gmb15]. The AUT and dummy chip assembly with molding encapsulation is shown in Fig. 5.2d [22].

An unencapsulated AUT and dummy chip assembly is manufactured by following only the first step of the above-mentioned process. In this case, the AUT and dummy chip are placed at a distance of approximately 1.1 mm and three wedge-wedge Au wirebonds (GSG configuration) are formed between the AUT and dummy chip. The wirebond diameter and the loop height are kept same as before. This leads to a wirebond loop length of around $\lambda_0/2$ (i.e., 1.23 mm) at 122 GHz with air as medium.

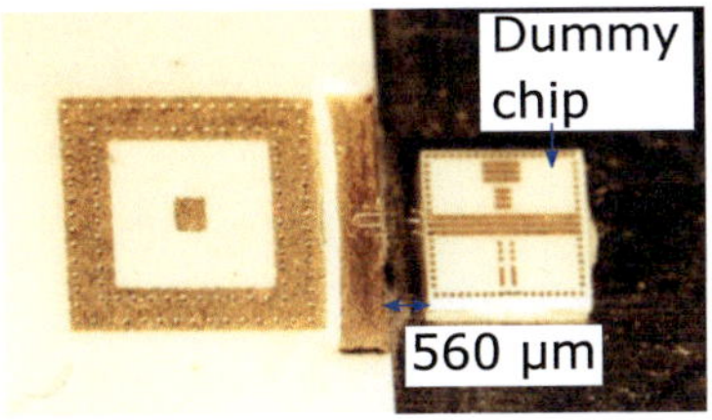

(a) mm-wave interconnect between the AUT and dummy chip

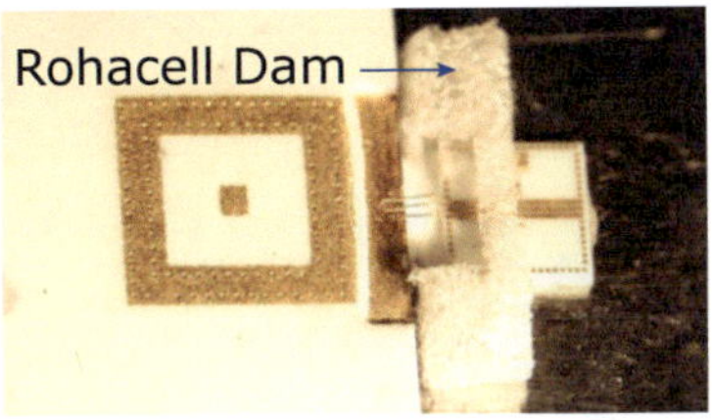

(b) Cavity formation by using a Rohacell structure

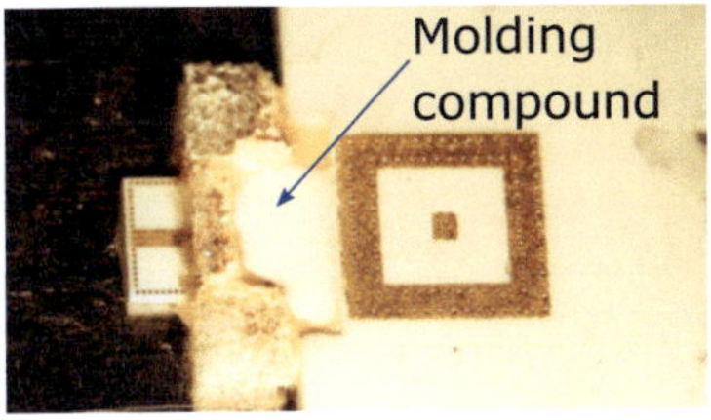

(c) Molding encapsulation before curing

(d) Molding encapsulation after curing

Figure 5.2. Building an AUT and dummy chip assembly with molding encapsulation [22]

Measurement and Simulation Results

The measurement of the encapsulated and unencapsulated assemblies of the AUT and dummy chip are carried out by using a GSG probe with 100 μm pitch. The measurements are preceeded by a gain calibration performed using a WR-6 standard rectangular horn and an SOL calibration performed using a standard calibration substrate. The simulation model of the encapsulated and unencapsulated assemblies are shown in Fig. 5.3a and Fig. 5.4a, respectively [22]. The following simplifications are used in both simulation models. First, the size of the AUT substrate and metal plate supporting the AUT and dummy chip are much smaller in the simulation than in reality. Second, the wirebonds are modeled using a standard spline geometry (available in the CST software) and the position of the maximum loop height is assumed to be in the middle of the wirebonds. In reality, since the wirebonds are manufactured manually, there is always a slight variation in the loop profile of each wirebond interconnect. Third, the simulated wirebonds are terminated ideally on the bond pads and the fine mechanical influences of the bonding tool on the wirebond geometry, such as bond foot length, front radius and back radius [Coo] are not taken into consideration. The simulations are performed using the CST MWS transient solver and the boundary conditions used in the simulation include an electric boundary beneath the metal plate and farfield boundaries on the remaining sides.

Unencapsulated Assembly of an AUT and a Dummy Chip

The measured and simulated RC curves are shown in Fig. 5.3b [22]. Both curves show a good agreement between 118 and 140 GHz. The measured RC at 122 GHz is −25.9 dB. The measured RC is less than −10 dB between 118.3 to 135.8 GHz. Fig. 5.3c shows the measured and simulated peak antenna gain versus frequency curves [22]. The maximum gain observed in the measurement and simulation is 10.7 dBi at 122.9 GHz and 9.5 dBi at 121 GHz, respectively. Further, in both curves, the peak antenna gain decreases at around 130 GHz (e.g., the peak antenna gain is 3.5 dBi in the measurement and 4.5 dBi in the simulation at 130 GHz) and then increases again at higher frequencies. The measured peak antenna gain is greater than 8 dBi between 116.6 and 126.2 GHz. The simulated total antenna efficiency in this frequency range varies between 66% and 77%.

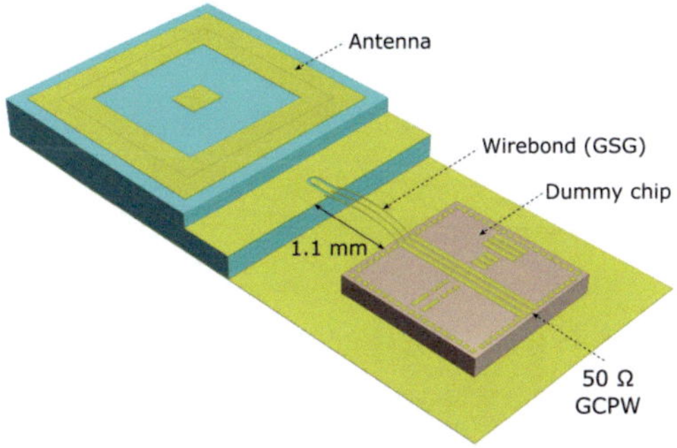

(a) Simulation model

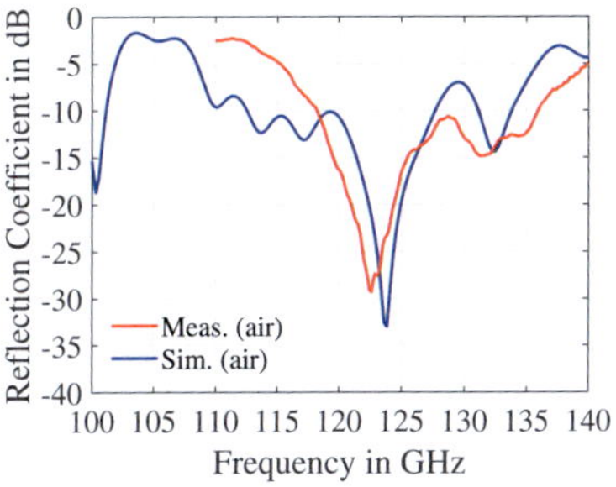

(b) RC

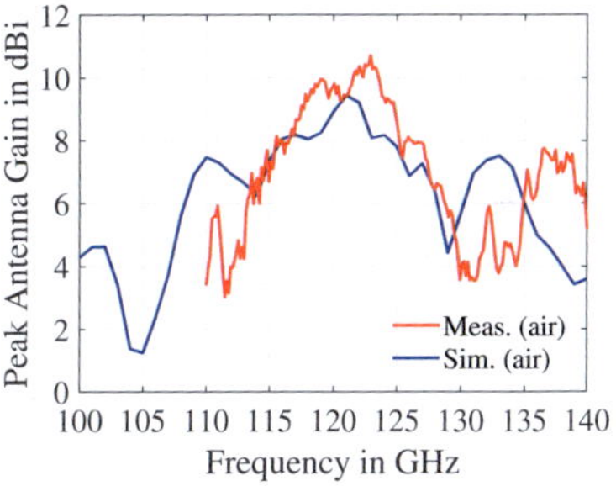

(c) Peak antenna gain

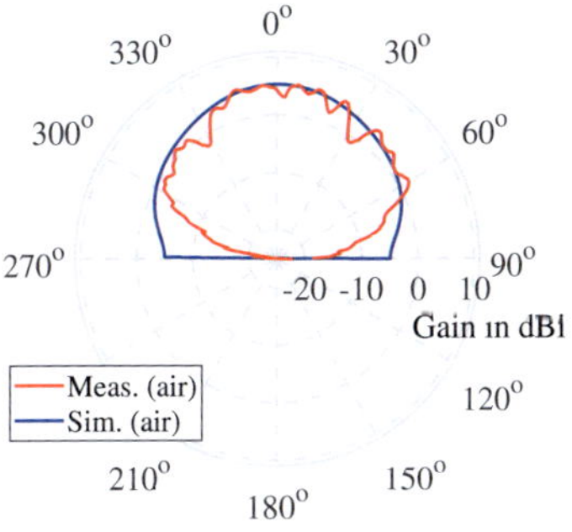

(d) Farfield radiation pattern in H-plane at 122 GHz

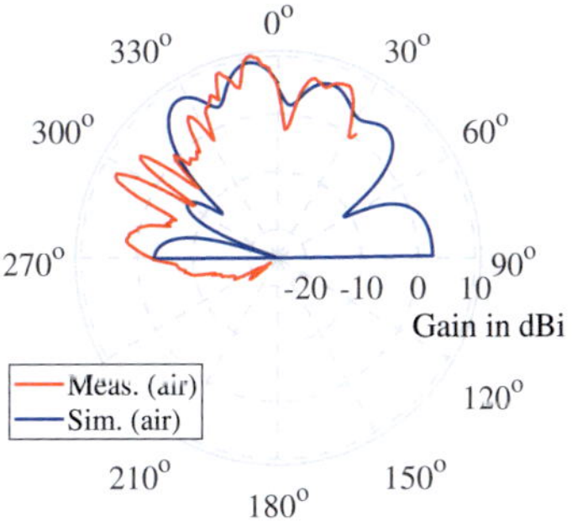

(e) Farfield radiation pattern in E-plane at 122 GHz

Figure 5.3: An unencapsulated AUT and dummy chip assembly [22]

The relative bandwidth over which the RC is less than -10 dB and the peak antenna gain is greater than 8 dBi is 6.5% (i.e., 118.3 to 126.2 GHz). The measured and simulated farfield radiation patterns at 122 GHz in H- and E-plane are shown in Fig. 5.3d and Fig. 5.3e, respectively [22]. In H-plane, the measured and simulated main lobes show a good agreement. In E-plane, both measurement and simulation show that the main lobe is tilted by approximately 8° counter-clockwise from the boresight direction. This tilt could have occurred due to the influence of the $\lambda_0/2$ wirebond interconnects. The antenna gain measured along the boresight direction is 4.3 dBi and the measured peak antenna gain is 10.4 dBi.

Encapsulated Assembly of an AUT and a Dummy Chip

The measured and simulated RC curves are shown in Fig. 5.4b [22]. The measured RC is less than -10 dB between 112 to 135.2 GHz, whereas the simulated RC is less than -10 dB between 107 to 132 GHz. As mentioned previously, the presence of voids in the molding encapsulation can shift its ε_r, which may be responsible for the frequency shift observed between the measured and simulated RC curves. Fig. 5.4c shows the measured and simulated peak antenna gain versus frequency curves [22]. The magnitude of the measured curve is 1 to 2 dB higher than the simulated curve, which could be attributed to the fact that the AUT substrate used in the measurement is much larger than in the simulation. The measured peak antenna gain varies between 8 and 9.9 dBi in the frequency range of 113 to 125 GHz. The simulated total antenna efficiency in this range lies between 60% and 64%. The relative bandwidth over which the measured RC is less than -10 dB and the measured peak antenna gain is greater than 8 dBi is 10.1%, i.e., 3.6% higher than the unencapsulated AUT and dummy chip assembly. The measured and simulated farfield radiation patterns at 122 GHz in H- and E-plane are shown in Fig. 5.4d and Fig. 5.4e, respectively [22].

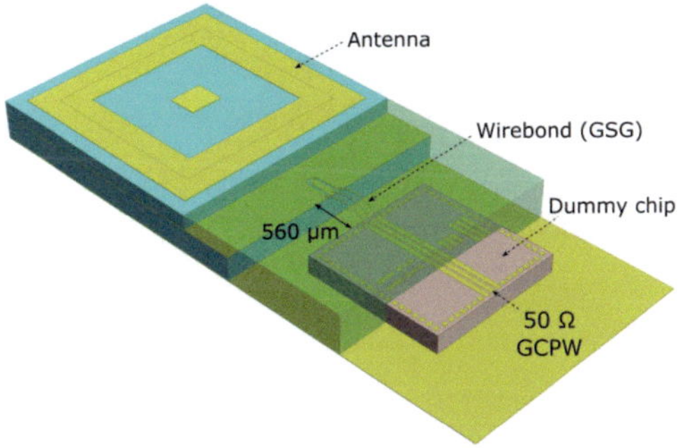

(a) Simulation model

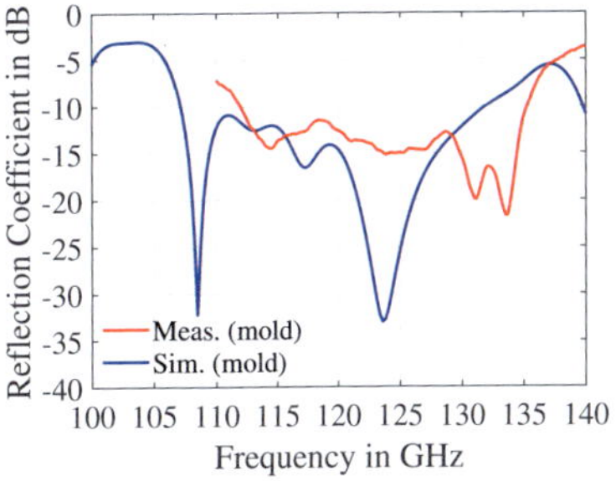

(b) RC

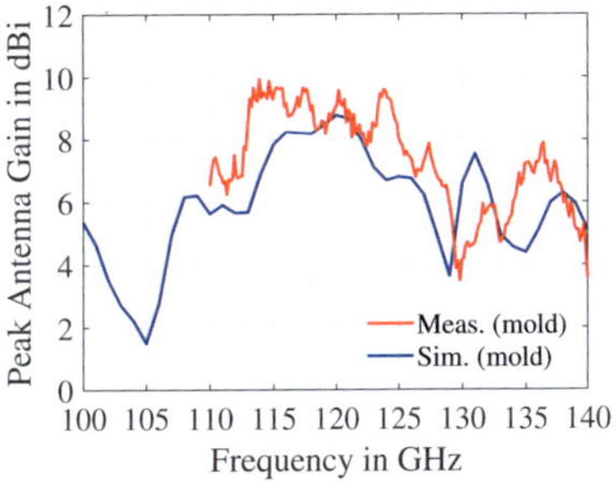

(c) Peak antenna gain

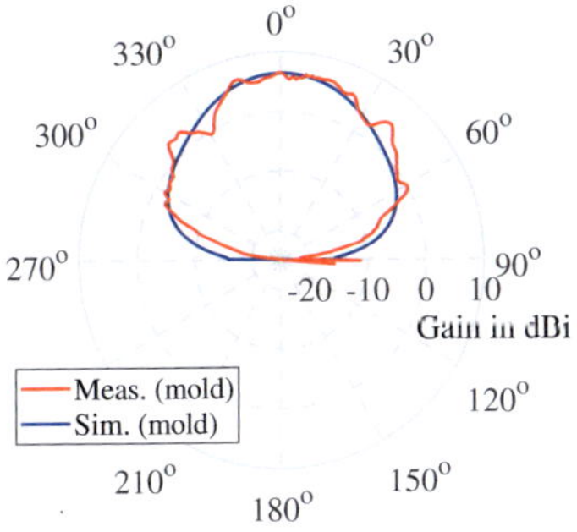

(d) Farfield radiation pattern in H-plane at 122 GHz

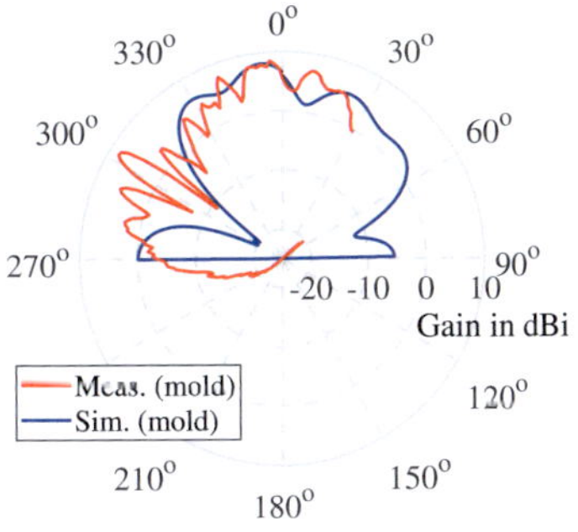

(e) Farfield radiation pattern in E-plane at 122 GHz

Figure 5.4: An encapsulated AUT and dummy chip assembly [22]

It is observed that in both H- and E-plane, the measured and simulated farfield radiation patterns are in good agreement with each other (barring the pattern enlargement in E-plane at around 300° due to the probe influence). The antenna gain measured along the boresight direction is 5.3 dBi. The peak antenna gain is measured as 8.5 dBi at an angle of 4° counter-clockwise from the boresight direction. The main lobe tilt is visible in both measurement and simulation. Therefore, it could be associated with the molded wirebond interconnects. Table 5.2 shows a comparison of the measured characteristics of the encapsulated and unencapsulated AUT and dummy chip assemblies. Two observations are made from this table. First, the relative bandwidth (i.e., the frequency range in which the measured RC is less than −10 dB and the measured peak antenna gain is greater than 8 dBi) is 3.5% higher with the encapsulated assembly because the impedance discontinuity caused by the mm-wave wirebond interconnects is relatively lower in the molding compound ($\varepsilon_r = 3.185$) than in air. Second, the measured peak antenna gain of the encapsulated assembly is 0.6 to 0.8 dB lower than the unencapsulated assembly due to the dielectric loss of the molding compound (i.e., $\tan\delta = 0.023$).

AUT - dummy chip assembly	Frequency in GHz		RC in dB		Peak Gain in dBi		BW in %
	Min	Max	Min	Max	Min	Max	
Air	118.3	126.2	-29.3	-10	7.4	10.7	6.5%
Polytec TC 430-T	113	125	-15	-11.5	8	9.9	10.1%

Table 5.2: A comparison of the measured characteristics of the encapsulated and unencapsulated assemblies of the AUT and dummy chip

5.2 QFN-Type Low Frequency Package Base in LTCC

The LTCC-based horizontal-configuration SiP concept shown in section 2.3.2 requires an LFPB for the following purposes. First, it acts as a low-frequency electrical interface between the MMIC and a PCB. Second, it enables thermal management of the MMIC and third, it provides mechanical support for the MMIC and mm-wave module assembly. A prototype of the LFPB used in this work is shown in Fig. 5.5 [10]. The LFPB is manufactured using DuPont 951 LTCC substrate, which is suitable for applications up to 35 GHz only. The electrical, thermal and mechanical properties of this substrate are given in Table 2.2. The LFPB dimensions are 8.8 mm×8.8 mm×0.4 mm. The top surface of the LFPB, on which the MMIC and mm-wave module are intended to be placed, is shown in Fig. 5.5a and the bottom surface of the LFPB, which is intended to be soldered to a PCB, is shown in Fig. 5.5b. The layout of the LFPB is similar to a QFN package. It consists of two large metal pads (5.5 mm×5.5 mm) in the middle, which are connected through dozens of thermal vias. Further, metal pads of size 230 µm×300 µm and pitch 500 µm are placed around the periphery of the bottom surface and they are connected to metal pads and conducting traces on the top surface by means of vias with a diameter of 180 µm. The LFPB is manufactured using standard LTCC manufacturing process, in which the conducting traces and vias are realized through standard screen printing and via punching processes. Therefore, the minimum line/space width and via diameter are restricted to 100 µm and 180 µm, respectively. A second variant of the LFPB is manufactured using Amkor GCS71 LTCC substrate, whose properties are given in Table 2.2. The only difference between the two LFPB variants is that an open air cavity is manufactured on the top surface of the second LFPB variant, which facilitates the molding encapsulation process. The following two subsections demonstrate the influence of the LFPB and molding encapsulation on the thermal and electrical performance of the horizontal-configuration SiP.

(a) Top view (b) Bottom view

Figure 5.5: A prototype of LTCC low frequency package base (LFPB) [10] © 2018 IEEE

5.2.1 Electrical Performance of a Horizontal-Configuration SiP

SiP I: mm-Wave Module with 2×2 Via-Fence Patch Arrays

The horizontal-configuration SiP concept is implemented using two different mm-wave modules. The first mm-wave module shown in Fig. 5.6a, consists of two 2×2 VF patch arrays, which serve as the Tx and Rx array of an MMIC transceiver, and their corresponding SL network-to-GCPW feeds. The design, simulation and measurement results of a single, unencapsulated 2×2 VF patch array has been shown in section 4.1.3. The Tx and Rx array of the mm-wave module differs from the previously shown single unencapsulated array in the following two aspects. First, the GCPW pads of the Tx and Rx array on ML 3 have a line and slot width of 60 μm and 41 μm, respectively, which provides an impedance of 50 Ω, if the GCPW line is encapsulated with Polytec TC 430-T (The GCPW terminal pads of the antenna array shown in section 4.1.3 provide an impedance of 50 Ω without an encapsulation). Second, the Rx array includes a large SL bend, which is required due to the Tx and Rx pad layout of the selected MMIC. The SiP simulation model is shown in Fig. 5.6b. It includes the aforementioned mm-wave module and an MMIC model, whose dimensions, layer stack and pad layout on the top ML resembles the MMIC used in this work. The mm-wave module and MMIC are interconnected by means of $\lambda_g/2$ GSG wirebonds on both Tx and Rx side and the assembly is encapsulated with Polytec TC 430-T.

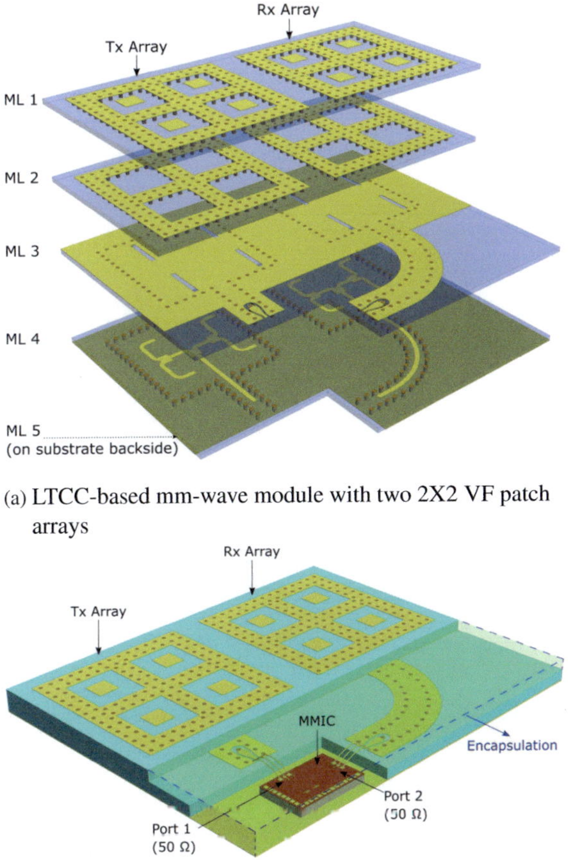

(a) LTCC-based mm-wave module with two 2X2 VF patch arrays

(b) Simulation model

Figure 5.6: LTCC-based horizontal-configuration SiP with Tx/Rx aperture-coupled 2×2 VF patch arrays and their SL network-to-GCPW feeds [10] © 2018 IEEE

The simulation uses two 50 Ω waveguide ports set up on the Tx and Rx GSG pads of the MMIC model (see Fig. 5.6b) and farfield boundary conditions on all sides. The simulation is performed using the CST MWS transient solver.

The corresponding simulation results are shown in Fig. 5.7. In terms of RC (see Fig. 5.7a), for the Tx array, S_{11} is less than -10 dB between 118 and 131 GHz and for the Rx array, S_{22} is less than -10 dB between 114 and 128.5 GHz. In contrast, a single, unencapsulated 2×2 VF patch array shows an RC less than -10 dB between 116.5 and 130.5 GHz (see section 4.1.3). Therefore, it is concluded that the molding encapsulation, large SL bend and $\lambda_g/2$ GSG wirebonds do not significantly affect the impedance bandwidth of the Tx and Rx array. Fig. 5.7b shows the coupling between the Tx and Rx array. The coupling over the entire frequency range of 110 to 150 GHz is less than -28 dB and the coupling in the frequency range of the MMIC is less than -35 dB. The coupling between the Tx and Rx array is very low, in spite of the fact that the Tx and Rx array are separated by only 650 µm (i.e., $0.25\lambda_0$ at 122 GHz). Hence, a high suppression of the surface waves is achieved. Fig. 5.7c shows the realized gain of the Tx and Rx array along the boresight direction and direction of the peak radiation. In both arrays, the farfield radiation pattern shows a minor tilt from the boresight direction. This tilt could be attributed to the large substrate size and the Tx and Rx array position on the substrate. The difference between the peak and boresight antenna gain is less than 0.7 dB over the entire frequency range. The maximum value of peak gain observed for the Tx and Rx array are 10.92 dBi and 10.74 dBi, respectively, at 125 GHz. In contrast, a single 2×2 VF patch array shows a simulated peak antenna gain of 12 dBi at 125 GHz (see section 4.1.3). This implies that the Tx array gain decreases by 1.08 dB due to the influence of the molding encapsulation and GSG wirebonds and the Rx array gain shows an additional loss of 0.18 dB due to the SL bend.

SiP II: mm-Wave Module with 6x6 Mushroom Electromagnetic Bandgap Antennas

The second mm-wave module used for building a horizontal-configuration SiP is shown in Fig. 5.8a [24]. This module consists of two 6×6 MEBG antennas (which serve as the Tx and Rx antennas of the MMIC transceiver) and their

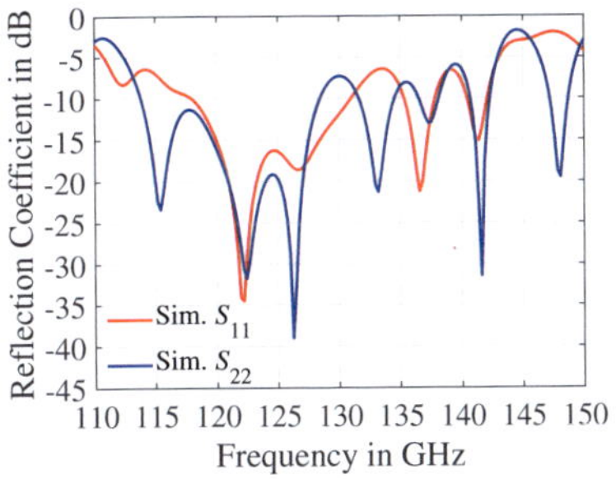

(a) Simulated RC of the Tx and Rx array

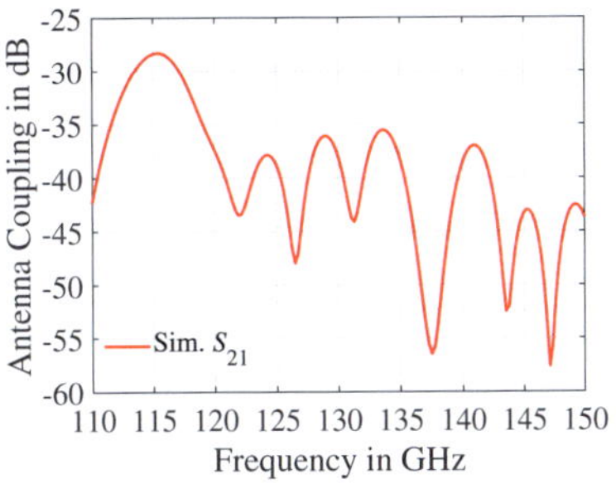

(b) Simulated coupling between the Tx and Rx array

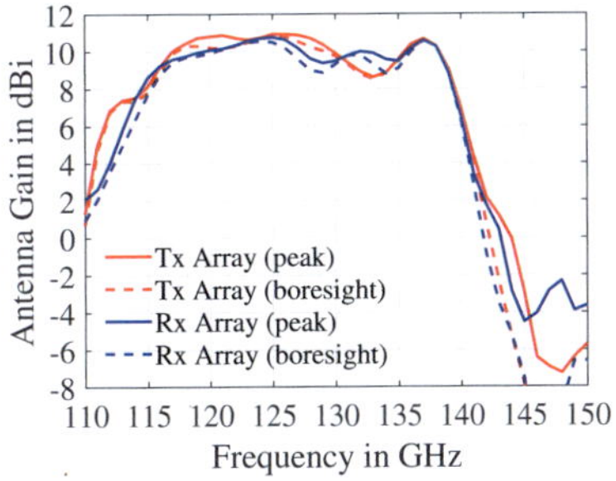

(c) Simulated peak and boresight gain of the Tx and Rx array

Figure 5.7: Simulation results of LTCC horizontal-configuration SiP I

corresponding SL-to-GCPW feeds. The design, simulation and measurement results of a single, unencapsulated 6×6 MEBG antenna with an SL-to-GCPW feed has been shown in section 4.2.4. Similar to SiP I, the GCPW terminal pads (on ML 2) of the Tx and Rx antennas provide an impedance of 50 Ω with Polytec TC 430-T as encapsulation. Also, the Rx antenna includes a large SL bend due to the MMIC pad layout. The corresponding SiP II simulation model is shown in Fig. 5.8b [24]. Besides the mm-wave module, the constituents of the SiP II simulation model are same as the SiP I. The simulation is performed using two 50 Ω waveguide ports shown in Fig. 5.8b. The boundary conditions and solver setup used in the simulation are also the same. The resultant simulation plots are shown in Fig. 5.9 [24]. In terms of RC (see Fig. 5.9a), S_{11} and S_{22} are less than -10 dB from 110 to 130 GHz. A single, unencapsulated 6×6 MEBG antenna shows an RC less than -10 dB between 112 and 134 GHz (see section 4.2.4). Therefore, the impedance bandwidth of the Tx and Rx antennas of SiP II is practically unaffected by the molding encapsulation, $\lambda_g/2$ GSG wirebonds and SL bend. In terms of antenna coupling (see Fig. 5.9b), the coupling between the Tx and Rx antennas is less than -32 dB between 110 and 125 GHz and less than -29 dB in the MMIC frequency range of 119.3 to 125.8 GHz. Further, the coupling increases between 125 and 140 GHz. The maximum value of mutual coupling is around -19 dB at 130 GHz. Note that the MEBG unit cell suppresses the surface waves only up to 123.5 GHz (see section 4.2.2) and therefore, the undesired surface waves propogate beyond 125 GHz, which result in an increased coupling between the Tx and Rx antennas. The edge-to-edge distance between the Tx and Rx antennas is 1.23 mm. Fig. 5.9c shows the realized gain of the Tx and Rx antennas along the boresight direction and the direction of peak radiation. In the MMIC frequency range, the peak antenna gain coincides with the boresight antenna gain and the main lobes of the Tx and Rx antennas do not show any tilt. The maximum value of antenna gain observed for the Tx and Rx antennas is 10.36 dBi and 10.15 dBi, respectively at 125 GHz. In contrast, a single, unencapsulated 6×6 MEBG antenna shows a maximum simulated antenna gain of 12 dBi at around 123 GHz (see section 4.2.4). This implies that the Tx antenna gain decreases by 1.64 dB due to the combined influence of the molding encapsulation and the GSG wirebonds. Further, the Rx array gain shows an additional loss of 0.21 dB due to the SL bend. Table 5.3 shows a summary of the simulated characteristics of SiP I and SiP II in the MMIC frequency range of 119.3 to 125.8 GHz.

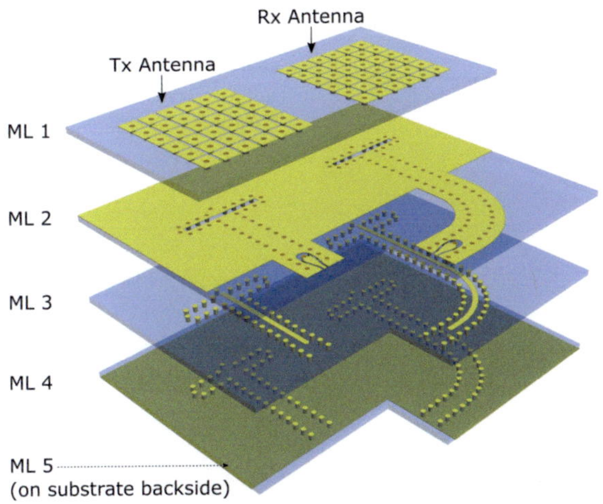

(a) mm-wave module with two 6X6 MEBG antennas

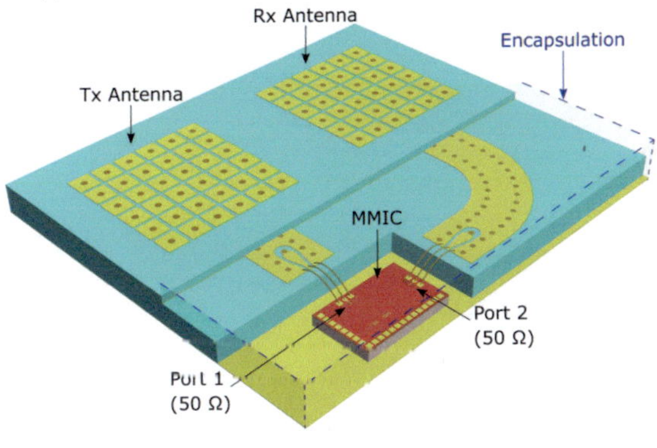

(b) Simulation model

Figure 5.8: LTCC-based horizontal-configuration SiP with Tx/Rx aperture-coupled 6×6 MEBG antennas and their SL-to-GCPW feeds [24]

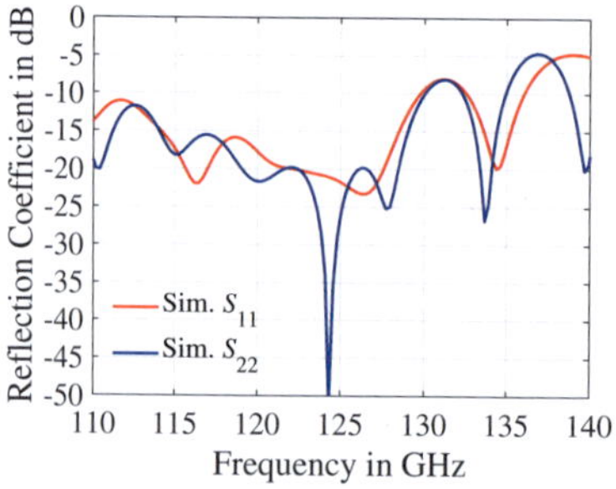

(a) Simulated RC of the Tx and Rx array

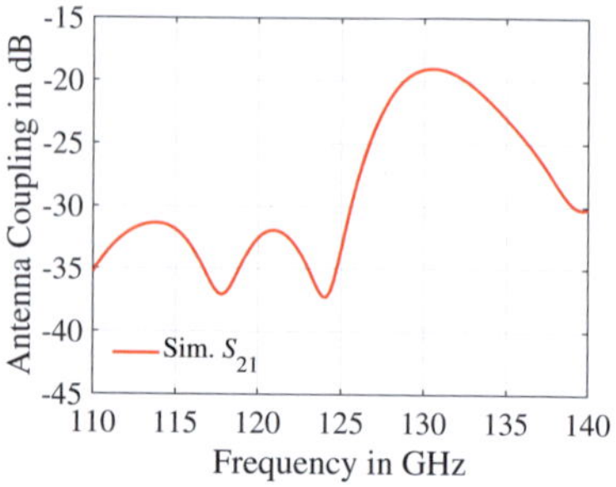

(b) Simulated coupling between the Tx and Rx array

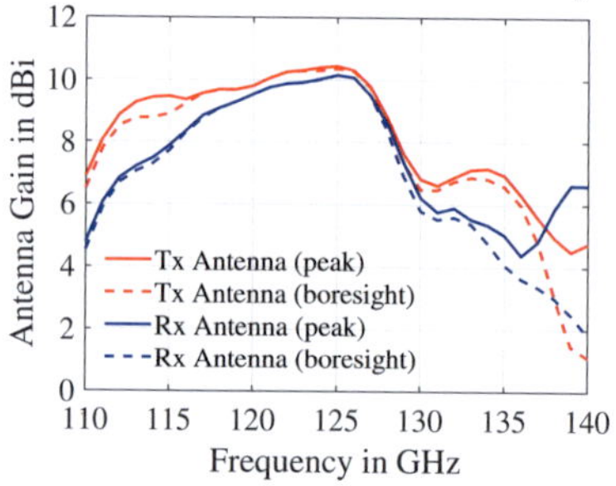

(c) Simulated peak and boresight gain of Tx and Rx antennas

Figure 5.9: Simulation results of LTCC horizontal-configuration SiP II [24]

	RC in dB		Antenna Coupling in dB		Boresight Antenna Gain in dBi	
	Min	Max	Min	Max	Min	Max
SiP I (Tx)	-34.5	-12.7	-45.1	-35.7	10.2	10.8
SiP I (Rx)	-31.7	-13.8	-45.1	-35.7	9.9	10.5
SiP II (Tx)	-23.3	-16.4	-40.1	-32.0	9.7	10.4
SiP II (Rx)	-50	-20.7	-40.1	-32.0	9.3	10.2

Table 5.3: A summary of the simulated characteristics of LTCC-based horizontal-configuration SiP I and SiP II in the MMIC frequency range (119.3 to 125.8 GHz)

5.2.2 Thermal Performance of a Horizontal-Configuration SiP

A thermal simulation is carried out in order to determine the influence of the horizontal-configuration SiP on the heat flow within the SiP and consequently on the maximum operating temperature of the MMIC. The thermal simulation model is shown in Fig. 5.10a [10]. It includes the wirebonded assembly of the MMIC and the mm-wave module (same as shown in Fig. 5.8b). The assembly is placed on a DuPont 951 substrate-based LFPB and the MMIC-to-LFPB (baseband) wirebond interconnects are modeled. Note that the MMIC is placed on the large metal pad of the LFPB, which is backed by dozens of thermal vias. The SiP, barring the radiating elements of the mm-wave module, is encapsulated with Polytec TC 430-T. The LFPB is mounted on a baseband PCB model of size 35 mm×35 mm×1 mm, which consists of four layers of flame retardant (FR)-4 substrate and four layers of Cu metallization. The thermal simulation is performed using the CST MPhysics studio thermal steady state solver. In the simulation, the MMIC with a power consumption of 380 mW acts as the heat source, the ambient temperature is set up as 21 °C and the phenomena of conduction and convection are taken into account. The phenomenon of conduction relies upon the thermal conductivity of the various materials, which are shown in Table 5.4. On the other hand, the phenomenon of convection relies upon the heat transfer coefficient of the various thermal surfaces. A thermal surface is a surface, which is in contact with the surrounding air and therefore supports natural convection. The thermal surfaces in this simulation are represented by the following interfaces: mm-wave module-to-air, LFPB-to-air, PCB-to-air, encapsulant-to-air and metal traces (on the mm-wave module, LFPB and PCB)-to-air. The heat transfer coefficient for free convection of gases typically varies between 2 and 25 W/m^2K [IDBL07]. In the simulation, the heat transfer coefficient of the first four thermal surfaces is assumed to be 5 W/m^2K and the heat transfer coefficient of the thermal surface involving metal traces is assumed to be 10 W/m^2K. The resultant temperature versus distance plot is shown in Fig. 5.10b [10]. In Fig. 5.10b, the horizontal axis shows the distance along the z-direction of the SiP-PCB assembly (see Fig. 5.10a) and the vertical axis shows the temperature variation along this distance. It is observed that a maximum temperature of 52 °C is recorded on the MMIC, which is located at $z = 0$. On either side of the MMIC, the temperature gradually decreases and approaches the ambient temperature. The minimum and maximum junction temperature specified for the MMIC used in this work are −50 °C and 150 °C, respectively [Gmb18]. Therefore, based on the thermal simulation result, it is concluded that the LFPB and molding encapsulation lead to an optimum thermal performance of the selected MMIC.

Material	Thermal conductivity (W/mK)
Si (MMIC)	148
SiO_2 (MMIC)	1.4
FR4 (PCB)	0.2
Cu (PCB)	401
DuPont 9K7 (mm-wave module)	4.6
DuPont 951 (LFPB)	3.3
Au (mm-wave module & LFPB)	314
Polytec TC 430-T (Encapsulant)	0.7

Table 5.4: Thermal conductivity values of various materials used in the thermal simulation

5.3 Summary

In this chapter, a near-hermetic packaging approach is selected over a hermetic packaging approach due to its low cost and ease of implementation. A near-hermetic package is implemented by means of either a glob-top or cavity-filled encapsulation concept. Among these two concepts, a cavity-filled encapsula-

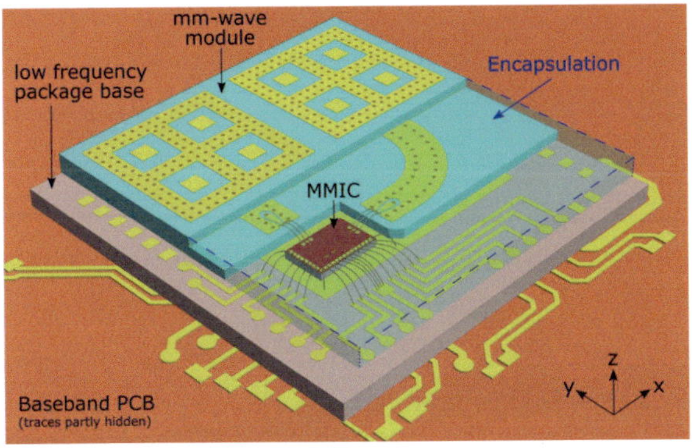

(a) Thermal simulation model

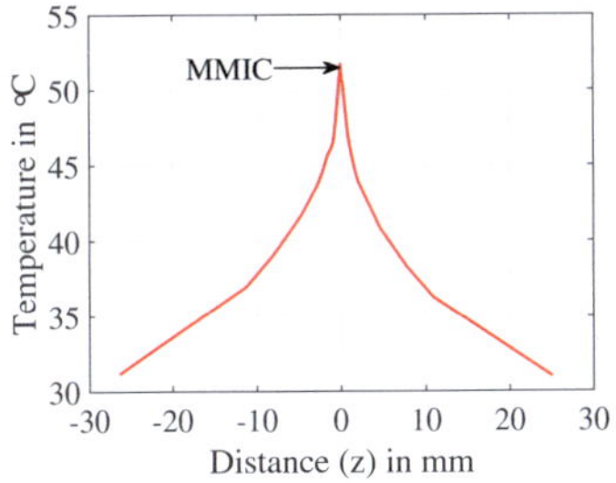

(b) Simulated temperature profile over the SiP height

Figure 5.10: Thermal simulation of LTCC-based horizontal-configuration SiP I surface mounted on a PCB [10] © 2018 IEEE

tion concept is preferred since it leads to a well-defined volume of the SiP encapsulant. As a result, the encapsulated SiP can be precisely modeled. Subsequently, the model can be used to perform EM and thermal simulations in order to determine the electrical and thermal performance of the encapsulated SiP. After determining the encapsulation process, the criteria used for selecting a molding compound as an encapsulant are presented. These criteria include thermomechanical properties (CTE, elasticity modulus), thermal properties (thermal conductivity, maximum operating temperature), physical properties (viscosity, adhesion, ease of storage and handling) and dielectric properties (ε_r, $\tan\delta$ in the frequency range of interest). Based on these criteria, an epoxy resin (Polytec TC 430-T) is selected as the package encapsulant. The main advantages of Polytec TC 430-T include the ability to withstand thermal stress, good thermal conductivity, a high maximum operating temperature, easy storage, suitability for manual application, availability of dielectric characteristics (from 110 to 170 GHz) and a relatively low dielectric loss in the frequency range of interest ($\tan\delta = 0.023$ at 122 GHz). The influence of the molding encapsulation is determined in practice by building a proof of concept, in which two assemblies are built using an AUT (i.e., a VF, SP antenna with an SL-to-GCPW feed shown in section 4.1.2) and a dummy chip (containing a 50 Ω GCPW). In the first assembly, the AUT and dummy chip are connected by three $\lambda_0/2$ GSG wirebonds (using air as medium). In the second assembly, the AUT and dummy chip are connected by three $\lambda_g/2$ GSG wirebonds and the assembly, barring the AUT's top radiating surface, is encapsulated with Polytec TC 430-T. As per the measurement result, the unencapsulated assembly shows an RC less than −10 dB and a peak antenna gain between 7.44 and 10.7 dBi from 118.3 to 126.2 GHz (i.e., 6.5% relative bandwidth). On the other hand, the encapsulated assembly shows an RC less than −11.5 dB and a peak antenna gain between 8 and 9.9 dBi from 113 to 125 GHz (i.e., 10.1% relative bandwidth). Therefore, the molding encapsulation helps in achieving a larger impedance bandwidth. However, it leads to a slight reduction in the peak antenna gain due to its dielectric loss. Thereafter, an LFPB made up of a low frequency LTCC substrate (DuPont 951) is introduced. At this point, all components required for building a horizontal-configuration SiP have been individually described in this work. Next, two horizontal-configuration SiPs are modeled, namely, SiP I and SiP II. These two models mainly differ in terms of the mm-wave module employed. The mm-wave module of SiP I includes two 2×2 VF patch arrays, whereas the mm-wave module of SiP II includes two 6×6 MEBG antennas. First, the electrical performances of the SiP I and SiP II are investigated through EM simulations. As per the simulation result of SiP I in the MMIC frequency range, the RC of the Tx and Rx array is less than −12.7 dB and −13.8 dB, respectively. The mutual coupling of the Tx and Rx array is less than −35.7 dB. The boresight antenna gain of the Tx and Rx array varies from 10.2 to 10.8 dBi and 9.9 to 10.5 dBi, respectively. As per the simulation result of SiP II in the MMIC frequency range, the RC of the Tx and Rx antenna is less than −16.4 dB and −20.7 dB, respectively. The mutual coupling of the Tx and Rx antenna is less than −32 dB. The boresight antenna gain of the Tx and Rx antenna varies from 9.7 to 10.4 dBi and 9.3 to 10.2 dBi, respectively. In both SiP I and SiP II, the Rx antenna gain is around 0.2 dB lower than the Tx antenna gain due to the large SL bend included on the Rx side. The peak antenna gain of SiP I is 1.2 to 1.5 dB lower than the peak antenna gain of a single, unencapsulated 2×2 VF patch array. The peak antenna gain of SiP II is 1.6 to 1.8 dB lower than the peak antenna gain of a single, unencapsulated 6×6 MEBG antenna. Finally, a thermal simulation is performed to determine the maximum operating temperature of the selected MMIC. The thermal simulation model includes the SiP I, LFPB and a baseband PCB model. As per the simulation result, for an ambient temperature of 21 °C and an MMIC power consumption of 380 mW, the maximum MMIC temperature is observed to be 52 °C. The simulated temperature of the MMIC is optimum, since it lies approximately in the middle of the minimum and maximum operating temperatures specified in the MMIC datasheet.

6 122 GHz FMCW Radar SiPs in LTCC Technology

In this chapter, two SiPs, namely, LTCC SiP I and LTCC SiP II are built by integrating a 122 GHz radar MMIC with two different mm-wave modules and two different LFPBs, respectively. Both SiPs employ a horizontal-configuration SiP approach, which was introduced in section 2.3.2. In both SiPs, a mm-wave interconnect between the MMIC and the mm-wave module is realized by means of three GSG-configuration $\lambda_g/2$ wirebonds with self-matching property. Further, the DC and baseband interconnects between the MMIC and LFPB are realized through conventional wirebonds without any matching network. An epoxy-based molding compound (Polytec TC 430-T) is used to encapsulate the MMIC and the wirebond interconnects. Both LTCC SiPs are surface mounted on two identical baseband PCBs, which are plugged into a microcontroller module, thus realizing two ready-to-use FMCW radar sensors at 122 GHz. The functionality of both FMCW radar sensors is verified by measuring the distance of a stationary target. In both cases, the radar performance is benchmarked by measuring the same target distance with a reference 122 GHz FMCW radar sensor, in which a commercially available radar SiP is surface mounted on an identical baseband PCB. The commercial radar SiP is provided by Silicon Radar GmbH. The benchmarking of LTCC SiPs is done in a fair manner as indicated by the following points. First, the 122 GHz radar MMICs used in the LTCC SiP I, LTCC SiP II and commercial SiP have the same specifications. Second, the three SiPs are surface mounted on identical baseband PCBs in order to build three individual FMCW radar sensors. Third, while measuring the target distance with each of these FMCW radar sensors, the Tx signal parameters, the target specifications and the signal processing parameters are kept the same.

6.1 122 GHz Radar MMIC

The 122 GHz radar MMIC used in this work is provided by Silicon Radar GmbH. It is fabricated in 0.13 μm SiGe BiCMOS technology and its dimensions are 1.475 mm×0.95 mm×0.164 mm. A micrograph of the top layer of the radar MMIC is shown in Fig. 6.1 [10]. At its core lies a VCO with four tuning voltage inputs (ranging from 0 to 3.3 V) and an output frequency range of 119.3 to 125.8 GHz. For signal transmission, the VCO signal is doubled, amplified by a power amplifier and then fed to an off-chip Tx antenna (or Tx array). For signal reception, the signal received by an off-chip Rx antenna (or Rx array) is fed into the MMIC, where the signal is first amplified by an LNA and subsequently down-converted to an IF signal by using a homodyne in-phase and quadrature (IQ) mixer. The I- and Q-channel IF signals are in differential configuration. The MMIC also consists of a $1/32$ frequency divider, which is combined with the VCO tuning inputs and an external phase locked loop (PLL) in order to operate the radar in an FMCW mode. The radar MMIC requires a supply voltage (V_{cc}) of 3.3 V and it has a power consumption of 380 mW [Gmb18]. The mm-wave and baseband I/O pads of the radar MMIC, which are used to form wirebond interconnects with the mm-wave module and the LFPB, respectively, are indicated in Fig. 6.1 [10]. In baseband, a total of 16 I/O pads (including two ground pads) are used to form standard wirebond interconnects with the LFPB (i.e., without using any impedance matching technique). In the mm-wave frequency range, two sets of GSG-configuration pads (marked as Tx and Rx pads in Fig. 6.1) are used to form self-matched $\lambda_g/2$ GSG wirebond interconnects with the Tx and Rx antennas (or arrays) of the mm-wave module. Each GSG pad has an output impedance of 50 Ω in air and a pad-pitch of 100 μm.

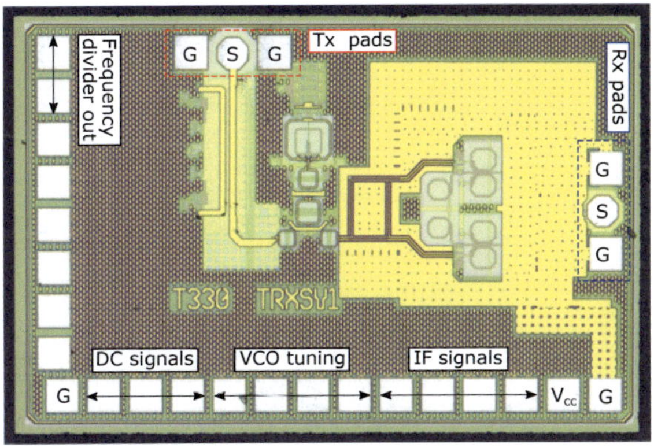

Figure 6.1: A micrograph of 122 GHz radar MMIC from Silicon Radar GmbH [10] © 2018 IEEE

6.2 LTCC SiP I

6.2.1 mm-Wave Module with 2×2 VF Patch Arrays

The LTCC SiP I is built using a mm-wave module, which consists of two 2×2 VF patch arrays. These arrays serve as the Tx and Rx array of the selected radar MMIC. Each array is fed by an SL network-to-GCPW signal transition. The multilayered structure of the mm-wave module is described in section 5.2.1. Fig. 6.2 shows a prototype of the mm-wave module [10]. The module has an overall size of 8 mm×6.5 mm and a thickness of approximately 420 µm. The bottom left corner of the module is diced in an L-shape (The dimensions of the cut portion are 3 mm×1.5 mm) in order to accommodate the radar MMIC. The manufacturing and dicing of the mm-wave module demands extreme precision, since the Tx/Rx GSG pads of the radar MMIC should be accurately aligned with the Tx/Rx GCPW terminal pads of the mm-wave module. Additionally, the edge-to-edge distance between the GSG pads of the radar MMIC and the GCPW terminal pads of the mm-wave module should be approximately 570 µm on both Tx and Rx sides. The former condition ensures that the GSG-configuration wirebonds are parallel to each other.

The latter condition is necessary to obtain wirebonds with a length of $\lambda_g/2$ in Polytec TC 430-T at 122 GHz. The Tx and Rx 2×2 VF patch arrays of the mm-wave module are measured without using the molding encapsulation. As mentioned in section 5.2.1, the Tx and Rx GCPW terminal pads of the mm-wave module are designed to have an impedance of 50 Ω in Polytec TC 430-T as required by the horizontal-configuration SiP concept. Further, the Tx and Rx GCPW terminal pads have an impedance of around 58 Ω in air. Note that the measurements shown in this section are made in air, since the Tx and Rx GCPW terminal pads cannot be contacted by a GSG probe if they are encapsulated with Polytec TC 430-T. The difference between the two impedance values is small and therefore, it does not lead to a significant difference in terms of the measurement results. The measurements are done with a GSG probe of 100 μm pitch. The array measurements are preceeded by a gain calibration performed using a standard WR-6 rectangular horn antenna and an SOL calibration performed using a standard calibration substrate. While measuring the Tx array (Rx array), the GCPW terminal pads of the Rx array (Tx array) are left open instead of being terminated by a 50 Ω matched load (unlike the real scenario). This should have practically no influence on the measurement results because as per the SiP I simulation results shown in section 5.2.1, the mutual coupling between the Tx and Rx array is less than −28 dB between 110 and 140 GHz and less than −35 dB in the frequency range of the radar MMIC. The simulation results of the mm-wave module are obtained by using only one waveguide port at a time (i.e., either for the Tx or Rx array) and farfield boundary conditions are applied on all sides. Both measurement and simulation results are normalized to 50 Ω. The FMCW radar sensors demonstrated in this work are operated between 121 and 125 GHz. Therefore, the values of the measured RC, peak and boresight antenna gain are specifically observed in this frequency range.

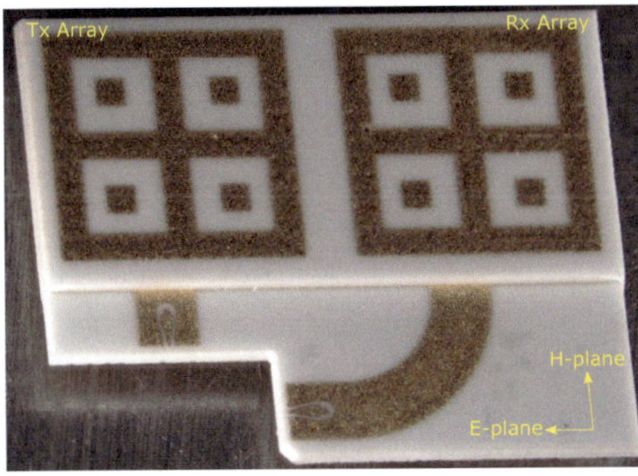

Figure 6.2: LTCC mm-wave module including 122 GHz Tx/Rx 2×2 VF patch arrays [10] © 2018 IEEE

Measurement and Simulation Results of Tx Array

Fig. 6.3 shows the measurement and simulation results of the Tx 2×2 VF patch array [10]. In terms of RC (see Fig. 6.3a), the resonance peaks observed in the measured RC curve are located at 120.5, 133 and 139 GHz. The first measured resonance peak matches well with the simulation. But the second and third peaks are shifted to higher frequencies in comparison with the simulation. As a result, the measured −10 dB RC range is also slightly lower than in the simulation. This discrepancy may have occurred due to a variation in the shrinkage of the substrate height during the LTCC co-firing process. The measured RC of the Tx array is less than −10 dB between 118.7 and 126.5 GHz. In the FMCW radar frequency range of 121 to 125 GHz, the measured RC of the Tx array varies between -12.8 and −24.6 dB. Fig. 6.3b shows the measured and simulated antenna gain versus frequency curves along the boresight direction and along the direction of maximum radiation. In the measurement, the direction of maximum radiation shows a tilt of less than 10° with respect to the boresight direction. A good agreement is observed between the measured and simulated gain curves, barring a sharp dip observed in the measurement at around 117.5 GHz. Note that the measured impedance matching around 117.5 GHz is degraded in comparison with the simulation and therefore a lower gain around this frequency is understandable. The measured boresight antenna gain of the Tx array is greater than 10 dBi between 120.5 and 132.8 GHz and the measured peak antenna gain of the Tx array is greater than 10 dBi between 119.6 and 134.6 GHz. In the FMCW radar frequency range of 121 to 125 GHz, the measured boresight antenna gain of the Tx array varies between 10.8 and 11.6 dBi and the measured peak antenna gain of the Tx array varies between 11.1 and 12.4 dBi. The total simulated antenna efficiency in this frequency range varies between 71 and 80%.

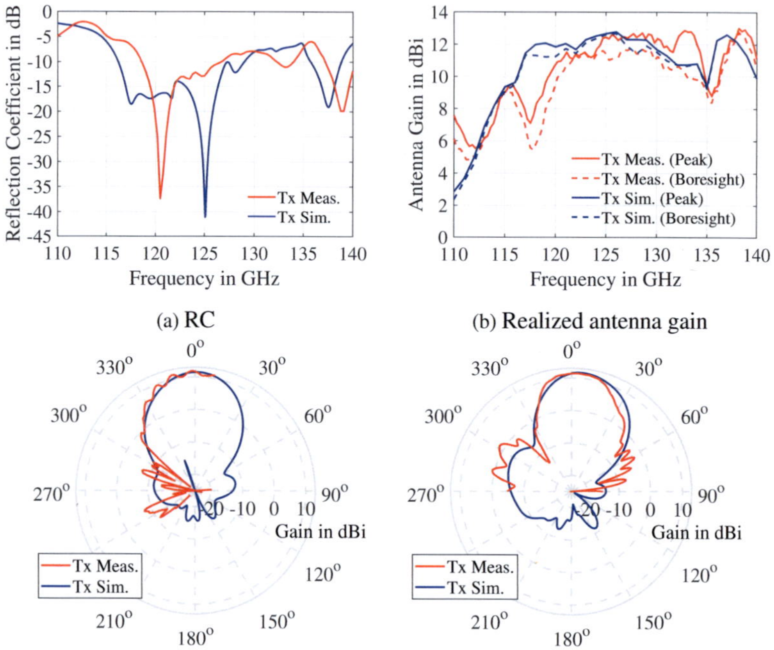

(a) RC

(b) Realized antenna gain

(c) Farfield radiation pattern in H-plane at 122 GHz

(d) Farfield radiation pattern in E-plane at 122 GHz

Figure 6.3: Measurement and simulation of Tx 2×2 VF patch array [10] © 2018 IEEE

The measured and simulated 122 GHz farfield radiation patterns of the Tx array in H- and E-plane are shown in Fig. 6.3c and Fig. 6.3d, respectively. The shape of the measured and simulated radiation patterns are in good agreement with each other. At 122 GHz, the measured boresight gain of the Tx array is 11.3 dBi and no tilt is observed in the measured main lobe.

Measurement and Simulation Results of Rx Array

Fig. 6.4 shows the measurement and simulation results of the Rx 2×2 VF patch array [10]. In terms of RC (see Fig. 6.4a), the resonance peaks observed in the measured RC curve are located at around 121, 130 and 136.4 GHz. Similarly to the Tx array, the first measured resonance peak matches well with the simulation, but the second and third peaks are shifted in frequency. Consequently, the measured −10 dB RC range is slightly lower than in the simulation. The measured RC of the Rx array is less than −10 dB between 118.4 and 125 GHz. In the FMCW radar frequency range of 121 to 125 GHz, the measured RC of the Rx array varies between -10.8 and −17.6 dB. Fig. 6.4b shows the measured and simulated antenna gain versus frequency curves along the boresight direction and along the direction of maximum radiation. Similarly to the Tx array, the direction of maximum radiation shows a tilt of less than 10° with respect to the boresight direction as per the measurement. In the frequency range of interest, a good agreement is observed between the measured and simulated gain curves. At frequencies around 118.5 and 132.5 GHz (i.e., outside the frequency range of interest), the measured gain is seen to be significantly lower than the corresponding simulated gain value. This discrepancy may be associated with the degradation in the measured impedance matching at these points. Moreover, a defect in manufacturing the large SL bend (e.g., misalignment of ground vias) might also be responsible for the degradation in gain at around 118.5 and 132.5 GHz. The measured boresight antenna gain of the Rx array is greater than 10 dBi between 122 and 127.7 GHz and the measured peak antenna gain of the Rx array is greater than 10 dBi between 121.1 and 128 GHz. In the FMCW radar frequency range of 121 to 125 GHz, the measured boresight antenna gain of the Rx array varies from 8.4 to 11.4 dBi and the measured peak antenna gain of the Rx array varies from 10.6 to 11.9 dBi. The total simulated antenna efficiency in this frequency range varies between 64 and 73%. Fig. 6.4c and Fig. 6.4d show the measured and simulated 122 GHz farfield radiation patterns of the Rx array in H- and E-plane, respectively. The shape of the measured and simulated radiation patterns are in good agreement with each other. At 122 GHz, the measured boresight gain of the Rx array is 10.9 dBi and the measured peak gain of the Rx array is 11.7 dBi (observed at an angle of 5° clockwise from the boresight direction in H-plane).

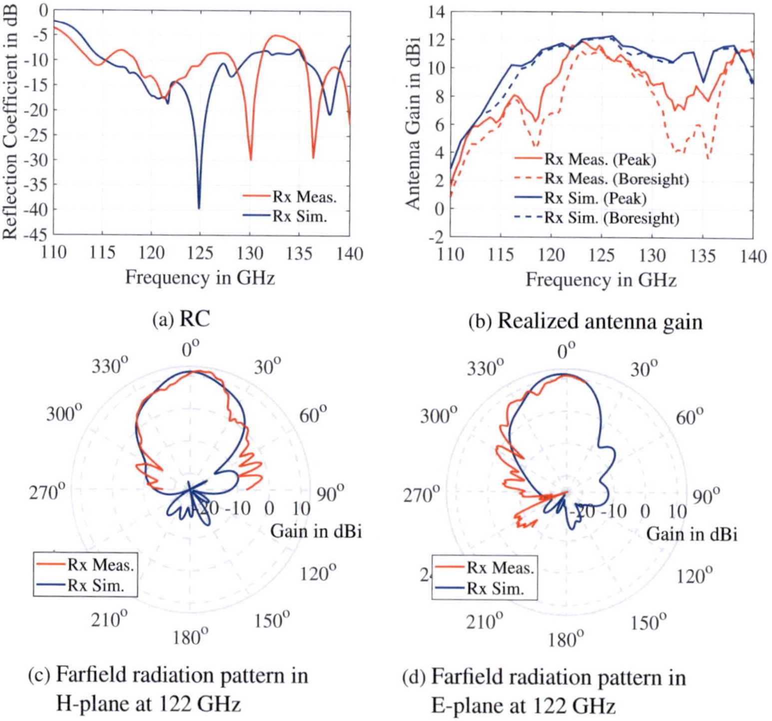

(a) RC

(b) Realized antenna gain

(c) Farfield radiation pattern in H-plane at 122 GHz

(d) Farfield radiation pattern in E-plane at 122 GHz

Figure 6.4: Measurement and simulation of Rx 2×2 VF patch array [10] © 2018 IEEE

A summary of the measured parameter values of the Tx and Rx 2×2 VF patch array in the FMCW radar frequency range of 121 to 125 GHz is shown in Table 6.1.

	RC in dB		Peak Array Gain in dBi		Boresight Array Gain in dBi		Sim. Array Efficiency in %
	Min	Max	Min	Max	Min	Max	
Tx Array	-24.6	-12.8	11.1	12.4	10.8	11.6	78
Rx Array	-17.6	-10.8	10.6	11.9	8.4	11.4	70

Table 6.1: A summary of the measured parameters of the Tx/Rx 2×2 VF patch array in the FMCW radar frequency range of 121 to 125 GHz

6.2.2 Manufacturing and Assembly of LTCC SiP I

The LTCC SiP I is manufactured by integrating the radar MMIC, the mm-wave module consisting of Tx/Rx 2×2 VF patch array and the DuPont 951 LFPB (shown in section 5.2). The steps involved in manufacturing the LTCC SiP I are shown in Fig. 6.5. The steps involved in manufacturing the LTCC SiP I are as follows.

1 The mm-wave module consisting of Tx/Rx 2×2 VF patch array is placed and attached to the top surface of the DuPont 951 LFPB.

2 The 122 GHz radar MMIC is placed and attached next to the mm-wave module such that the following two conditions are satisfied. First, the Tx/Rx GSG pads on the radar MMIC are aligned with the Tx/Rx GCPW terminal pads on the mm-wave module. Second, the edge-to-edge distance between the GSG pads on the radar MMIC and the GCPW terminal pads on the mm-wave module is ≈ 570 µm on both Tx and Rx sides. Third, the radar MMIC is in contact with the thermal pad of the LFPB. The placement and attachment of both the mm-wave module and the

radar MMIC is carried out using a flip-chip bonder. The radar MMIC is attached to the thermal pad of the LFPB by using Polytec TC 430-T, which is a thermally-conductive, electrically-insulating epoxy adhesive.

3 The DC and baseband wirebond interconnects are formed between the radar MMIC and LFPB. These wirebonds do not employ any form of impedance matching technique since the frequency of the signals routed through these wirebonds is less than 3 GHz. The interconnects are realized using an Au wirebond of 17.5 µm diameter and a wedge-shaped bonding tool.

4 The mm-wave wirebond interconnects are formed between the GSG pads of the radar MMIC and the GCPW terminal pads of the mm-wave module on both Tx and Rx sides. These interconnects are realized in wedge-wedge configuration by using an Au wirebond of 17.5 µm diameter. The wirebond loop height is kept ≈100 µm above the mm-wave module, which leads to an overall length of $\approx\lambda_g/2$ at 122 GHz in Polytec TC 430-T medium. All wirebonds are manufactured using a semi-automatic wirebonder. The wirebond interconnects between the radar MMIC, mm-wave module and LFPB are shown in Fig. 6.5a [10].

5 Since the DuPont 951 LFPB has no open-cavity, a Rohacell ($\varepsilon_r\approx1$) structure is glued on three sides of the LFPB in order to form a makeshift open-cavity. Thereafter, the open-cavity is filled with the selected epoxy-based molding compound (Polytec TC 430-T is prepared by mixing its components in a specific ratio, as prescribed in the manufacturer's data sheet [Gmb15]). The adhesive is gradually filled in order to avoid formation of any air bubbles during the molding encapsulation. Finally, the assembly is cured at a temperature of 150 °C for 15 min. The LTCC SiP I during an intermediate stage of the molding encapsulation is shown in Fig. 6.5b and the LTCC SiP I in its final form (post curing) is shown in Fig. 6.5c [10].

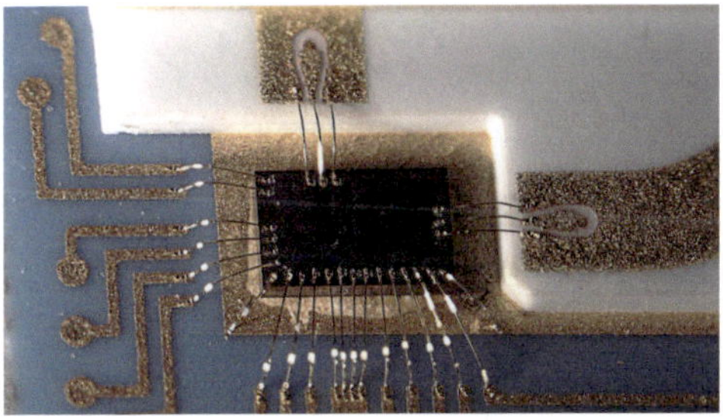

(a) Wirebond interconnects between the radar MMIC, mm-wave module and LFPB

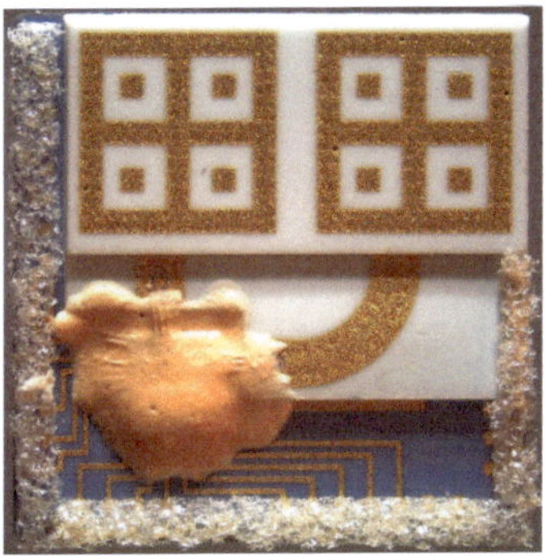

(b) An intermediate stage of the molding encapsulation process

(c) LTCC SiP I with molding encapsulation post curing

Figure 6.5: Manufacturing steps of LTCC SiP I [10] © 2018 IEEE

6.2.3 122 GHz FMCW Radar Range Measurement

The LTCC SiP I is surface mounted on a baseband module in order to realize a ready-to-use 122 GHz FMCW radar. The baseband module RAPID v3 is developed by Wellenzahl GmbH & Co. KG. It mainly consists of baseband filters, anti-aliasing filters, variable gain baseband amplifiers, two 12-bit analog-to-digital converters (ADCs) with a maximum sampling rate of 5 MHz, a PLL and a Cortex M4-based microcontroller with digital signal processing (DSP) capabilities [uSGCK]. Fig. 6.6a shows the LTCC SiP I surface mounted on a baseband PCB and Fig. 6.6b shows the 122 GHz FMCW radar in its entirety [10]. The FMCW radar is used to measure the distance of a standard stationary target, i.e., a trihedral corner reflector with a side length (a) of 42.43 mm. The effective area (A_{eff}) and the radar cross section (σ) of a trihedral corner reflector are given by Eq. 6.1 and Eq. 6.2, respectively.

$$A_{\text{eff}} = a^2/\sqrt{3} \tag{6.1}$$

$$\sigma = \frac{4\pi a^4}{3\lambda_0^2} \tag{6.2}$$

Using the above equations, the A_{eff} and σ of the trihedral corner reflector are calculated as 1039.4 mm^2 and 3.51 dBsm, respectively. The target is placed at a distance (R) of 1.15 m. The FMCW signal ramp generated for transmission has a start frequency (f_{start}) of 121 GHz, a stop frequency (f_{stop}) of 125 GHz, a bandwidth (B=$f_{\text{stop}} - f_{\text{start}}$) of 4 GHz, a sweep time (T_{sweep}) of 0.384 ms and sawtooth modulation. Multiple FMCW radar measurements are made for the same target distance. For every measurement, the received FMCW signal ramp is down-converted to differential I- and Q-channel IF signals, which are then digitized by the 12-bit ADCs. The resultant consists of 2048 IF data samples for each channel (i.e., I- and Q-channel), which are processed offline using MATLAB software. The steps involved in the offline processing of the IF data samples are as follows.

1 The 2048 I- and Q-channel IF data samples obtained from the multiple measurements of a fixed target distance are averaged.

2 The average I- and Q-channel IF data samples are combined to obtain the complex IF data samples.

3 A Hamming window function is applied to the resultant IF data samples, which reduces the side lobe level at the cost of an increase in the target range resolution.

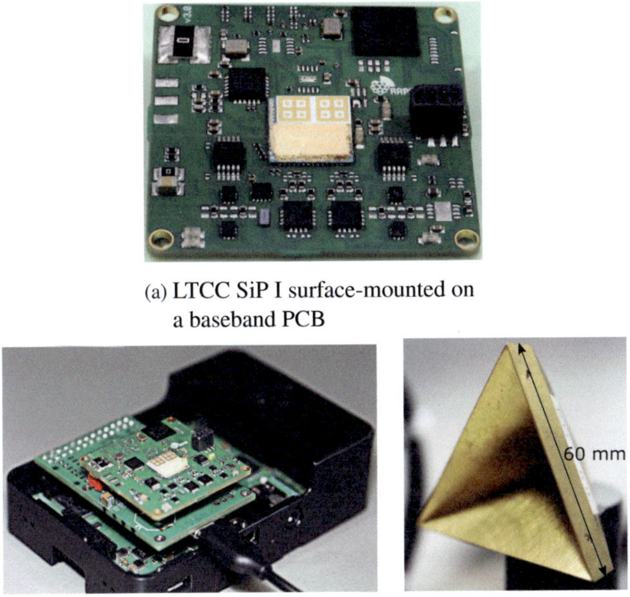

(a) LTCC SiP I surface-mounted on a baseband PCB

(b) 122 GHz FMCW radar sensor based on LTCC SiP I

(c) Radar target: Trihedral corner reflector

Figure 6.6: Functionality test of LTCC SiP I [10] © 2018 IEEE

4 The windowed signal is transformed into the frequency domain by performing a zeropadded FFT with a zeropadding factor of 32768. Consequently, the frequency corresponding to the target peak (f_{IF}) is determined.

5 The target distance is estimated by using Eq. 6.3.

$$R = \frac{c_0 T_{\mathrm{sweep}} f_{\mathrm{IF}}}{2B} \tag{6.3}$$

Table 6.2 shows a summary of the FMCW signal ramp parameters, target specifications and signal processing parameters. In order to benchmark the performance of the LTCC SiP I, the same target distance is measured by using a commercial 122 GHz radar SiP from Silicon Radar GmbH. In the commercial SiP, the 122 GHz radar MMIC (shown in Fig. 6.1) is integrated with a Tx and Rx antenna in a plastic molded QFN package. The key selling points of the LTCC SiP I in comparison to the commercial SiP are given below.

Parameter	Value
Start frequency (f_{start})	121 GHz
Stop frequency (f_{stop})	125 GHz
Sweep bandwidth (B)	4 GHz
Sweep time (T_{sweep})	0.384 ms
ADC samples	2048 (I/Q channel)
Target distance (R)	1.15 m
Zeropadded FFT factor	32768
Window function	Hamming

Table 6.2: FMCW radar measurement and signal processing parameters [10] © 2018 IEEE

1 The LTCC SiP I off-chip circuitry uses moderate values of minimum line and space width i.e., 60 μm and 40 μm, respectively. On the other hand, the commercial SiP is a modified version of a 122 GHz radar SiP shown in [BRG$^+$13], in which the off-chip circuitry includes extremely narrow line and space widths of approximately 20 μm. In addition, the circuitry is realized by using a relatively expensive, non-standard thin film process [BRG$^+$13]. The cost of the LTCC SiP is comparatively lower.

2 The Tx and Rx antenna gain of the LTCC SiP I is 2 to 3 dB higher than the commercial SiP. In addition, the footprint of both SiPs is nearly the same [Gmb18].

3 In the LTCC SiP, the multi-step cavity structure generated by the horizontal configuration SiP concept naturally facilitates a selective molding encapsulation. This is not true for the commercial SiP, since the antenna substrate height is much lower than the molding encapsulation.

The commercial radar SiP is surface mounted on an identical baseband module (i.e., the baseband module is the same as shown in Fig. 6.6b) in order to build a reference 122 GHz FMCW radar for target distance measurement. The FMCW signal ramp parameters, target specifications and signal processing parameters are kept the same in both cases. Therefore, a fair comparison between the two radar SiPs has been made. Fig. 6.7a shows the baseband IF signal measured by the 12-bit ADC over a time period of T_{sweep}=0.384 ms [10]. The 12-bit ADC requires a supply voltage of 3.3 V. Therefore, the IF signal amplitude is obtained as ADC Data $\times$ 3.3 V/4096 (A 12-bit ADC has 2^{12}=4096 quantization levels). As seen in the plot, the peak-to-peak IF signal amplitude measured with the LTCC SiP I is larger than the commercial SiP. This indicates that the Tx and Rx array gain of LTCC SiP I is higher than the Tx and Rx antenna gain of the commercial SiP. In the FMCW radar range (i.e., 121 to 125 GHz) the simulated Tx and Rx antenna gain of the commercial SiP varies approximately from 8 to 9 dBi and 8.5 to 10 dBi, respectively, as per the datasheet shown in [Gmb18]. On the other hand, the measured Tx and Rx array gain of the LTCC SiP I varies from 11.1 to 12.4 dBi and 10.6 to 11.9 dBi, respectively, in the same frequency range (as shown in Table 6.1). Therefore, it is justified to obtain a larger peak-to-peak IF signal amplitude using LTCC SiP I. Fig. 6.7b shows the IF spectrums obtained after performing the zeropadded FFT [10]. For both SiPs, barring the DC peak, the strongest peak is observed at a frequency of around 80 kHz, which corresponds to the target. Therefore, the target IF frequency is given by f_{IF}=80 kHz. Further, the peak amplitude of the target IF frequency obtained with LTCC SiP I is nearly 0.75 dB higher than the commercial SiP. Fig. 6.7c shows a zoom-in view of the target distance versus amplitude plot [10]. For both SiPs, the target peak is observed at a distance of nearly 1.15 m. The 3 dB range resolution of the FMCW radars built using the LTCC SiP I and the commercial SiP are measured as 65.6 mm and 67.9 mm, respectively. The theoretical value of range resolution with a rectangular window function is $c_0/2B$=37.5 mm.

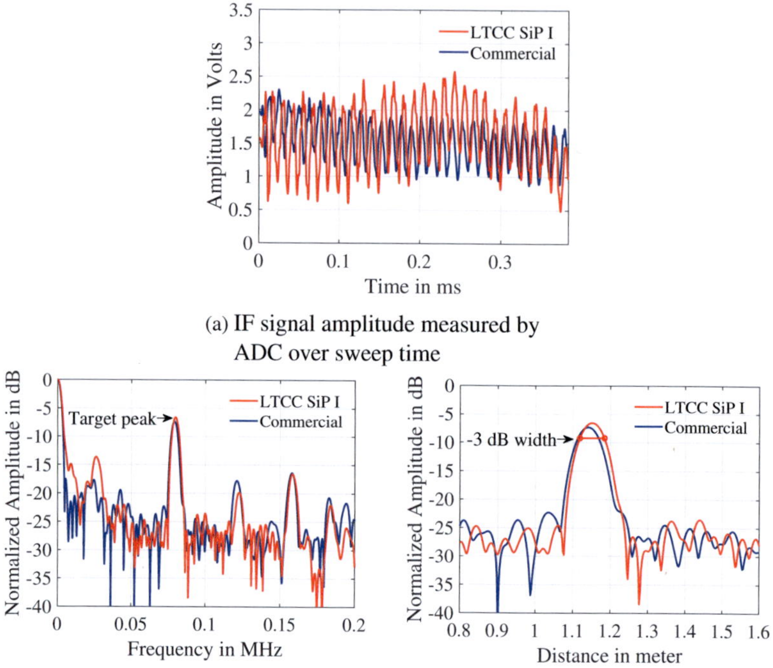

(a) IF signal amplitude measured by ADC over sweep time

(b) IF signal spectrum after zeropadded FFT

(c) Estimation of target range and range resolution

Figure 6.7: Measurement and benchmarking of 122 GHz FMCW radar sensor employing LTCC SiP I [10] © 2018 IEEE

6.3 LTCC SiP II

6.3.1 mm-Wave Module with 6×6 MEBG Antennas

The LTCC SiP II is built by using a mm-wave module, which consists of two 6×6 MEBG antennas, which serve as the Tx and Rx antenna for the selected radar MMIC. Each antenna is fed by an asymmetric SL-to-GCPW signal transition. The multilayered structure of this mm-wave module is described in section 5.2.1 and its prototype is shown in Fig. 6.8 [24]. The module has an overall size of 7 mm×5.9 mm and a thickness of $\approx$ 420 µm. Similarly to the 2×2 VF patch array-based mm-wave module shown in the previous section, a high degree of precision is required in manufacturing the Tx/Rx GCPW terminal pads as well as in dicing the mm-wave module. This ensures that the Tx/Rx GSG wirebond interconnects between the radar MMIC and the mm-wave module are parallel and have an overall length of $\lambda_g/2$ in Polytec TC 430-T as medium.

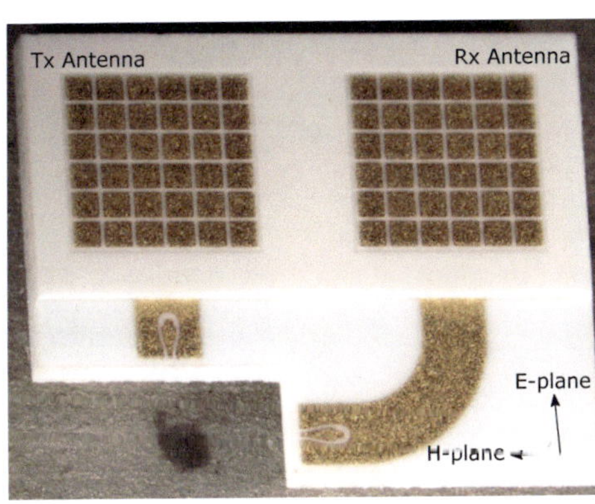

Figure 6.8: LTCC mm-wave module including 122 GHz Tx/Rx 6×6 MEBG antennas [24]

The Tx and Rx 6×6 MEBG antennas are measured one after the other using a GSG probe with 100 µm pitch. The measurement and simulation conditions are the same as described in section 6.2.1. As per the SiP II simulation result shown in section 5.2.2, the mutual coupling between the Tx and Rx antenna is less than −32 dB between 110 and 125 GHz and less than −19 dB between 125 and 140 GHz. Therefore, while measuring the Tx (Rx) antenna, the GCPW terminal pads of the Rx (Tx) antenna are left open without causing any significant change in the measured parameters. Once again, the values of the measured RC and the realized antenna gain of both Tx and Rx antennas are specifically observed between 121 and 125 GHz, since the mm-wave module is used to build a second FMCW radar sensor, which is operated in the same frequency range.

Measurement and Simulation Results of Tx Antenna

Fig. 6.9a shows the measured and simulated RC of the Tx antenna [24]. The first two resonance frequencies observed in the measured RC are located at 114 and 123.5 GHz, which are seen in the simulated RC as well. However, in the desired frequency range, the magnitude of the measured impedance matching is severely degraded in comparison with the simulation. As previously shown in section 4.2.4, simulative investigations and X-ray analysis of a single 6×6 MEBG antenna indicate that the antenna performance is significantly influenced by a layer misaligment of 60 µm. Therefore, the degraded measured impedance matching could be attributed to layer misalignment during the LTCC manufacturing process. The measured RC between 121 and 125 GHz varies from −6.1 to −9.5 dB. Further, the measured RC shows an improvement in the higher frequency range (i.e., between 135 and 150 GHz), which also indicates towards a layer misalignment. Fig. 6.9b shows the realized peak antenna gain versus frequency curves [24]. In this plot, Tx Meas. (I) indicates the measured peak antenna gain and Tx Meas. (II) indicates the deembedded peak antenna gain, which is obtained after deembedding the influence of the measured impedance mismatch. The latter is calculated by using the equation $\mathrm{RG}_{\mathrm{deembed}}=\mathrm{RG}_{\mathrm{meas}}/(1-|S_{11}|^2)$, where $\mathrm{RG}_{\mathrm{deembed}}$, $\mathrm{RG}_{\mathrm{meas}}$ and S_{11} denote the deembedded antenna gain, measured antenna gain and measured RC, respectively. The plot shows that the deembedded and simulated peak antenna gain curves are in agreement with each other. The measured peak antenna gain between 121 and 125 GHz varies from 8 to 10.5 dBi. The maximum and minimum gain values are observed at 123.5 and 125 GHz, respectively. The deembedded antenna gain in this frequency range varies from 10.3 to 12.1 dBi. Fig. 6.9c and Fig. 6.9d show the farfield radiation patterns in H- and E-plane, respectively, at 122 GHz [24]. The shape of the measured and deembedded farfield radiation patterns match with the simulated farfield radiation pattern in both planes. At 122 GHz, the measured and deembedded antenna gain along the boresight direction are 8.7 and 10.5 dBi, respectively.

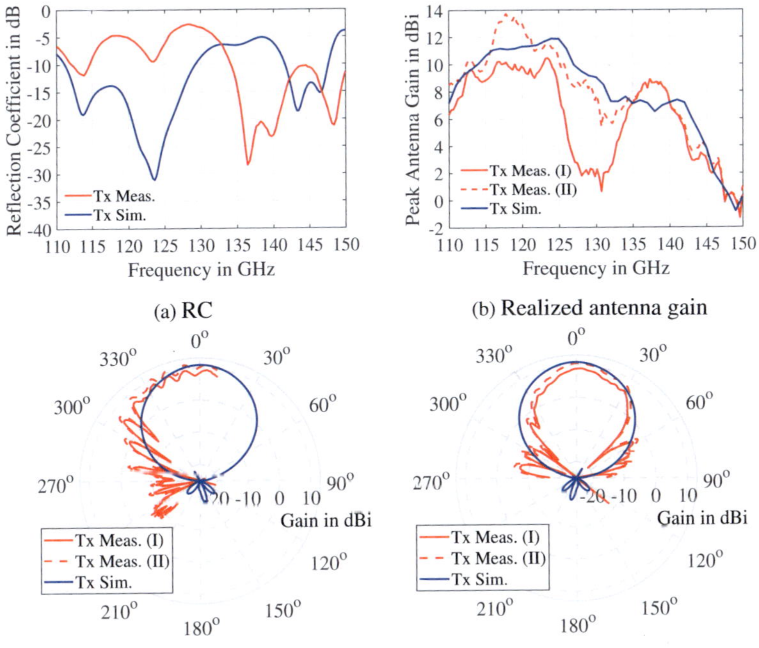

(a) RC

(b) Realized antenna gain

(c) Farfield radiation pattern in H-plane at 122 GHz

(d) Farfield radiation pattern in E-plane at 122 GHz

Figure 6.9: Measurement and simulation of Tx 6×6 MEBG antenna (Tx Meas. (I) and Tx Meas. (II) show measured and deembedded result, respectively) [24]

Measurement and Simulation Results of Rx Antenna

Fig. 6.10a shows the measured and simulated RC of the Rx antenna [24]. The first three resonance frequencies observed in the measurement are located at 114, 121 and 128 GHz. Among these, the first resonance frequency is visible in the simulation, whereas the second and third resonance frequencies are somewhat in agreement with the simulation. Similarly to the Tx antenna, the measured impedance matching at these frequencies show a significant degradation, in comparison with the simulation, which may have occurred due to layer misalignment during the LTCC manufacturing process. Between 121 and 125 GHz, the measured RC varies between −3.5 and −16.6 dB. Fig. 6.10b shows the peak antenna gain curves [24]. In this plot, the measured peak antenna gain is shown by Rx Meas. (I) and the deembedded peak antenna gain is shown by Rx Meas. (II). Rx Meas. (II) shows an improvement over Rx Meas. (I). The discrepancy between Rx Meas. (II) and Rx Sim. curves could be associated with the defects involved in manufacturing the large asymmetric SL bend on Rx side. Fig. 6.10c and Fig. 6.10d show the farfield radiation patterns in H- and E-plane, respectively, at 122 GHz [24]. The shape of the measured and deembedded farfield radiation patterns match with the simulated farfield radiation pattern in both planes. At 122 GHz, the measured and deembedded antenna gain along the boresight direction are 6.4 and 7.6 dBi, respectively.

6.3.2 122 GHz FMCW Radar Assembly and Measurement

An LTCC SiP II is manufactured using the mm-wave module with a Tx and Rx 6×6 MEBG antenna. The manufacturing process is the same as described in section 6.2.2, for LTCC SiP I. The only difference is that the LFPB used in this case is an open-cavity package, whose details are given in section 5.2. The open-cavity facilitates the molding encapsulation process and therefore no Rohacell make-shift structure is required while manufacturing the LTCC SiP II. Fig. 6.11a shows the baseband and mm-wave wirebond interconnects be-

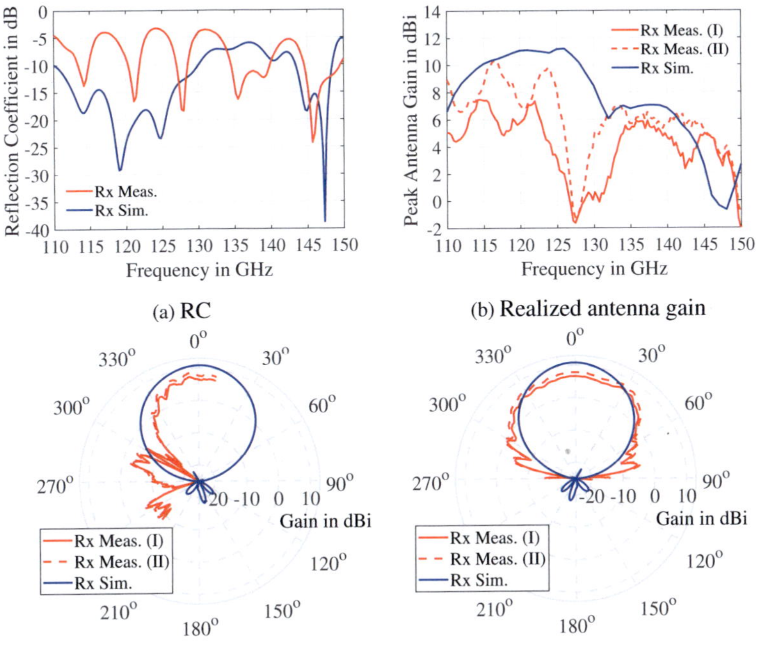

(a) RC

(b) Realized antenna gain

(c) Farfield radiation pattern in H-plane at 122 GHz

(d) Farfield radiation pattern in E-plane at 122 GHz

Figure 6.10: Measurement and simulation of Rx 6×6 MEBG antenna (Rx Meas. (I) and Rx Meas. (II) show measured and deembedded result, respectively) [24]

tween the radar MMIC, the mm-wave module and the LFPB [24]. Fig. 6.11b shows the LTCC SiP II with Polytec TC 430-T-based molding encapsulation post curing [24]. The LTCC SiP II is surface mounted on a baseband module in order to realize a second FMCW radar sensor at 122 GHz, as shown in Fig. 6.11c [24]. The baseband module is the same as the one used for LTCC SiP I in the previous section. The functionality of the FMCW radar is tested and benchmarked against the reference 122 GHz FMCW radar, which consists of the commercial radar SiP surface mounted on an identical baseband module.

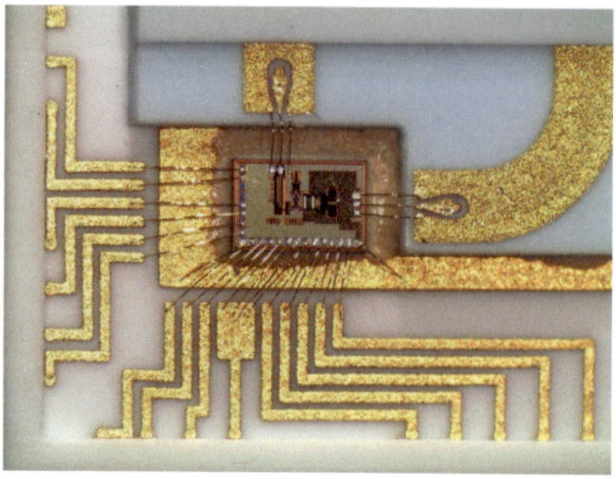

(a) Wirebond interconnects between the radar MMIC, mm-wave module and open-cavity LFPB

(b) LTCC SiP II with molding encapsulation post curing

(c) LTCC SiP II surface mounted on a baseband PCB

Figure 6.11: Manufacturing steps of LTCC SiP II [24]

The FMCW ramp parameters, target specifications and signal processing parameters are the same as shown in Table 6.2. The measured plots are shown in Fig. 6.12 [24]. Fig. 6.12a shows the baseband IF signal measured by the 12-bit ADC over a time period of T_{sweep}=0.384 ms. The peak-to-peak IF signal amplitude measured with the LTCC SiP II is smaller than the commercial SiP, since the measured Tx and Rx antenna gains are lower than the commercial SiP. Fig. 6.12b shows the IF spectrums after zeropadded FFT [24]. The target IF frequency obtained with both SiPs is f_{IF}=80 kHz. The peak amplitude of the target IF frequency obtained with LTCC SiP II is nearly 3 dB lower than the commercial SiP, which is due to the lower Tx/Rx antenna gain of the LTCC SiP II. Fig. 6.12c shows a zoom-in view of the target distance versus amplitude plot [24]. The target distance and 3 dB range resolution are measured as 1.15 m and 67.9 mm, respectively.

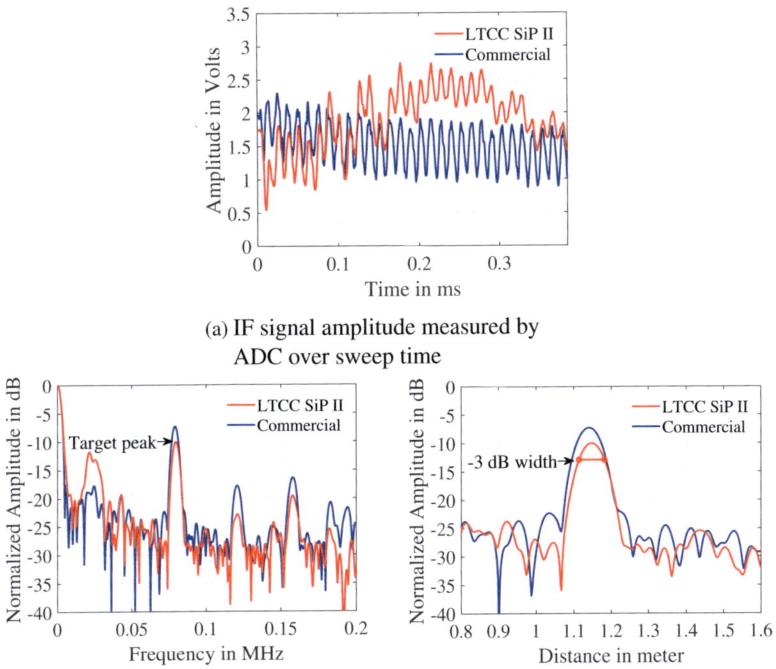

(a) IF signal amplitude measured by ADC over sweep time

(b) IF signal spectrum after zeropadded FFT

(c) Estimation of target range and range resolution

Figure 6.12: Measurement and benchmarking of 122 GHz FMCW radar sensor employing LTCC SiP II [24]

7 Conclusion

The tremendous progress made by semiconductor technologies combined with our quest to exploit vast amounts of unlicensed spectrum available at mm-wave frequencies has led to the realization of fully-integrated transceiver MMICs operating at frequencies beyond 100 GHz. Given the enormous potential to use these transceiver MMICs in a wide variety of radar and wireless communication based consumer applications, it becomes necessary to develop a mm-wave package, which acts as an electrical and mechanical interface between the MMIC and the outside world. In addition, it should provide protection, power and cooling to the MMIC. A waveguide-based split-block package, in spite of its high performance and long-term reliability, is considered impractical for this purpose due to its bulky nature and high cost. Moreover, the cost seldom goes down, not even for high-volume manufacturing. Therefore, a compact, robust and low-cost mm-wave SMP is required, which can be directly mounted on a standard PCB, in order to build a ready-to-use mm-wave system. In this work, an SiP approach has been selected for the implementation of a mm-wave SMP due to the advantages of its flexible antenna design, higher antenna radiation efficiency and lower cost. Among the available SiP technologies, the multilayer ceramic substrate technology is capable of realizing a multitude of high-quality passive devices, e.g., antenna, signal transition, filter and coupler, in the package itself, thus providing a wide range of configurations for a miniaturized 3D SiP design. Besides, a package based on a ceramic substrate guarantees system reliability under harsh environmental conditions, e.g., wide temperature range, humidity, corrosive gases and radiation. In this work, a multilayer ceramic technology named LTCC has been used. This technology has the following benefits. First, a variety of physical features including conducting patterns, vias, air cavities, stepped bonding planes and fluidic channels, can be integrated in a compact 3D LTCC module. Second, the LTCC technology is suitable for a heterogenous system integration, since the CTE of LTCC substrates lies close to the CTE of semiconductor materials, e.g., SiGe, GaAs and InP. Third, an LTCC-based SiP can efficiently dissipate the heat generated by an MMIC by using thermal vias in a standard LTCC package and fluidic channels in an advanced LTCC package. Fourth, the LTCC modules are highly resistant against thermal stress, mechanical stress, shock and vibration. Besides, they are lifelong hermetic. Until recently, LTCC-based passive components were demonstrated at frequencies below 100 GHz, mainly for three reasons. First, the LTCC substrates provided by well-reputed suppliers like DuPont, Heraeus, Kyocera, Ferro and ESL were characterized up to 10 GHz only. Second, the manufacturing resolution was limited due to the conventional thick film screen printing and via punching processes. Consequently, the minimum line/space width, via diameter and via pitch were restricted to a few hundred micrometers. Third, the manufacturing defects induced by the LTCC process, e.g., substrate shrinkage and layer misalignment, were relatively high. In the past few years, extensive research and development has been carried out, both in terms of materials and manufacturing techniques used in the LTCC technology, which has opened new avenues for manufacturing LTCC-based mm-wave components and systems beyond 100 GHz. The LTCC substrate selected in this work is DuPont 9K7, whose dielectric characteristics are available up to frequencies as high as 500 GHz. In addition, it has a stable and low loss tangent of around 0.001 up to and beyond 100 GHz. In terms of manufacturing, various fine-line patterning techniques have been recently made available, which have significantly improved the minimum possible feature size. One such technique, namely, laser ablation, is used in this work to realize a minimum line and space width of 60 μm and 40 μm, respectively. In addition, the micro-via technology has been used in this work to realize a minimum via radius and a corresponding via catchpad radius of 40 µm and 50 µm, respectively. The primary challenge involved in realizing an LTCC SiP at frequencies above 100 GHz is that, although the dimensional resolution and accuracy of LTCC has shown a tremendous improvement, it still lags behind the thin film technology. Additionally, a layer-to-layer misalignment of the order of tens of micrometers also exists. Taking the recent developments and the existing challenges into consideration, this work aims at developing an LTCC SiP for a SiGe bistatic radar MMIC, which operates between 119.3 and 125.8 GHz. The MMIC requires a supply voltage of 3.3 V and a power consumption of 380 mW. The main achievements of this work are elucidated below.

LTCC-based Horizontal-Configuration SiP Concept for mm-Wave Transceivers Beyond 100 GHz

An LTCC-based horizontal-configuration SiP concept for the integration of transceiver MMICs operating at frequencies up to and beyond 100 GHz is presented for the first time. The off-chip package circuitry is divided into two separate modules, namely, a mm-wave module and an LFPB. The mm-wave module is responsible for transmitting and receiving the mm-wave signal. It consists of one or more aperture-coupled antennas and a via-based SL-to-GCPW signal transition, feeding each of the antennas. The LFPB serves a threefold purpose. First, it helps in routing the low frequency signals (e.g., voltage supply and IF signal) between the MMIC and a standard baseband PCB. Second, it uses thermal vias to dissipate the heat generated by the MMIC. Third, it facilitates the SiP encapsulation. The LFPB and MMIC are interconnected through standard wirebonds (without using any matching technique), since the parasitic effects arising due to the wirebonds are almost negligible at frequencies up to a few gigahertz. On the other hand, the mm-wave module and the MMIC are interconnected through self-matched, half-wavelength wirebonds in GSG configuration. The SiP is partially encapsulated with an epoxy-based molding compound, Polytec TC 430-T, whose dielectric characteristics between 110 and 170 GHz are available. Moreover, Polytec TC 430-T shows a stable relative permittivity (ε_r) and a low loss tangent ($\tan\delta$) in this frequency range, e.g., at 122 GHz, the values of ε_r and $\tan\delta$ are 3.185 and 0.023, respectively. The molding compound encapsulates the mm-wave wirebond interconnects, which reduces the wirebond length by a factor of $1/\sqrt{\varepsilon_r}$. The strong points of this SiP concept are as follows. First, the SiP design is robust, since the typical LTCC manufacturing defects, e.g., shrinkage, layer misalignment and warpage are taken into consideration. Second, the SiP concept can be used up to a frequency of 150 GHz. The upper frequency limit of the proposed SiP is determined by the via-based SL-to-GCPW signal transition. Third, the SiP employs only standard assembly techniques. The length of the half-wavelength mm-wave wirebond interconnects varies from 634 to 423 µm in the frequency range of 100 to 150 GHz, respectively. Consequently, the wirebonds can be easily realized using a standard wirebonding tool. Further, a standard molding encapsulation is used for the SiP, which can be automated in an industrial environment as well. In terms of performance, the molding encapsulation has a minor influence on the radiation characteristics of the antenna. The thermal dissipation provided by the SiP is sufficient for MMICs with a power consumption of a few hundred milliwatts.

mm-Wave SL-to-GCPW Signal Transitions in LTCC Technology

A via-based SL-to-GCPW signal transition operating over a broad bandwidth of DC to 150 GHz has been demonstrated for the first time in LTCC technology. The signal transition is implemented by designing a controlled-impedance signal via. A signal via is first analyzed theoretically on the basis of its lumped element model, which leads to an approximate calculation of the via impedance (Z_{via}) for the given via dimensions, e.g., via radius and via height. Next, two variants of the via-based signal transition, namely, with air cavity (AC) and without air cavity (NC), are investigated by means of a parameter study, in which various parameters of the signal transition are varied one after the other and their influence is analyzed in time domain as well as frequency domain by observing the simulated impedance over time (i.e., time domain reflectometry) plots and the simulated S parameter plots, respectively. The S parameter measurement result of the signal transition with air cavity shows an RC less than -10 dB and a single-ended TC greater than -0.85 dB over a broad bandwidth of DC to 150 GHz. In contrast, the measurement of the signal transition without air cavity shows an RC less than -7.5 dB and a single-ended TC greater than -1.2 dB in the same frequency range. Hence, the role of an air cavity in mitigating the undesired parasitic effects at high frequencies is verified. The signal transition prototypes are also analyzed by means of X-ray microtomography, based on which their initial simulation models are modified. An excellent agreement is obtained between the measurement and the modified simulation results of the signal transition prototypes. An alternative approach based on EM coupling is also used to realize an SL-to-GCPW signal transition. This signal transition employs a ground plane aperture and a modified OC structure. The initial simulation result of the signal transition shows an RC less than -14 dB, a single-ended TC greater than -1 dB and a radiation loss of approximately 10% between 114.5 and 132 GHz. The X-ray microtomography of a prototype of the EM-coupled signal transition reveals that the signal conductors of the SL and GCPW are misaligned by approximately 40 µm and 50 µm along the transversal and longitudinal directions, respectively. The initial simulation model of the signal transition is modified on the basis of the X-ray analysis and a good agreement is achieved between the measurement and the modified simulation result. As per the S parameter measurement, the EM-coupled signal transition shows a single-ended TC greater than -2 dB from 120.5 to 135 GHz. In addition, the -10 dB RC bandwidth is shifted to higher frequencies. The degradation of the measured S parameters is associated with layer misalignment during the LTCC manufacturing process. Therefore, an improvement in the layer alignment accuracy is necessary to implement the proposed EM-coupled SL-to-GCPW signal transition at frequencies as high as 122 GHz.

mm-Wave Antennas in LTCC Technology

Various configurations of a via-fence (VF) and a Mushroom electromagnetic bandgap (MEBG)-based aperture-coupled antenna have been implemented in LTCC technology at 122 GHz. In the first antenna concept, a VF (i.e., a type of soft surface) is used to suppress the TM_0 surface wave mode and hence improve the farfield radiation pattern. Besides, the VF enhances the impedance bandwidth of the antenna as well. The VF-based antenna variants implemented in this work include a VF stacked-patch (SP) antenna with an MS-to-GCPW signal transition feed, a VF SP antenna with an SL-to-GCPW signal transition feed and a 2×2 VF patch array with an SL network-to-GCPW feed. A good agreement is achieved between the measurement and simulation results of all three antenna variants. The last among these variants (i.e., a 2×2 VF patch array with an SL network-to-GCPW feed) is used for the implementation of a horizontal-configuration LTCC SiP. In this array, the radiating elements are separated by a distance of 1.55 mm. The simulated loss incurred by the feed network, which consists of three quarter-wave matched SL power dividers, four 90° SL curved bends and an SL-to-GCPW via-based signal transition, is calculated as 1.8 dB. As per the measurement results of the array, the RC is less than −10 dB and the realized array gain varies between 10 and 12.85 dBi in the frequency range of interest. The simulated total antenna efficiency is greater than 70% in the same frequency range. The second antenna concept is based on a matrix of MEBG unit cells. The highlight of this antenna concept is that the antenna substrate requires a mere thickness of $0.04\lambda_0$ at 122 GHz, yet the antenna is capable of achieving a high bandwidth-efficiency product. The lumped element model of an MEBG unit cell resembles a composite right-/left-handed (CRLH) structure. This concept has been implemented for the first time in LTCC at such a high frequency. An MEBG unit cell has two key characteristics. First, its resonance frequencies occur in a non-harmonic ratio due to the nonlinear nature of its dispersion diagram, which leads to a large impedance bandwidth of the antenna. Second, it stops the surface wave propagation in specific frequency bands, known as electromagnetic bandgaps (EBGs). As a result, the radiation efficiency of the antenna increases. These two properties are optimized in the desired frequency range by simulating the dispersion diagram of an MEBG unit cell. The MEBG-based antenna variants implemented in this work include the following: a 4×4 MEBG antenna with an MS-to-GCPW signal transition feed; a 6×6 MEBG antenna with an MS-to-GCPW signal transition feed; a 4×4 MEBG antenna with an SL-to-GCPW signal transition feed; a 6×6 MEBG antenna with an SL-to-GCPW signal transition feed. The last among these antenna variants (i.e., 6×6 MEBG antenna with an SL-to-GCPW signal transition feed) is used for the implementation of a horizontal-configuration LTCC SiP. As per the measurement result, the antenna gain varies between 10 and 12.3 dBi in the desired frequency range. The buried substrate layers of the antenna are analyzed using X-ray microtomography, which indicates a layer misalignment of 60 µm or more. A simulative study also confirms that a layer misalignment is responsible for the discrepancies observed between the measured and initially simulated RC and realized gain versus frequency curve of the antenna. It is concluded that the performance of a 6×6 MEBG antenna is susceptible to the typical LTCC manufacturing tolerance of layer misalignment, whereas a 2×2 VF patch array proves to be robust enough to withstand the typical LTCC tolerances.

122 GHz FMCW Radar SiPs in LTCC Technology

Two 122 GHz LTCC-based radar SiPs are realized in this work namely, LTCC SiP I and LTCC SiP II. These two LTCC SiPs integrate a 122 GHz SiGe bistatic radar MMIC by implementing the proposed horizontal-configu ration SiP concept. The two SiPs mainly differ in terms of their mm-wave module. The mm-wave module used in LTCC SiP I consists of a Tx and an Rx 2×2 VF patch array and two SL network-to-GCPW signal transitions feeding each of the arrays. The mm-wave module used in LTCC SiP II consists of a Tx and an Rx 6×6 MEBG antenna and two SL-to-GCPW signal transitions feeding each of the antennas. The steps involved in manufacturing and encapsulating the LTCC SiPs are illustrated. The two molded LTCC SiPs are surface mounted on identical baseband PCB modules, thus realizing two ready-to-use FMCW radar sensors at 122 GHz. The functionality of both radar sensors is successfully verified by measuring the distance of a stationary target. Moreover, in both cases, the radar performance is benchmarked by measuring the same target distance with a commercial 122 GHz FMCW radar sensor. The bistatic radar MMIC and the baseband module used in the three radar sensors are identical. The FMCW signal ramp parameters, the target specifications and the signal processing parameters used in the three cases are also the same. Hence, a fair comparison is made between the two LTCC SiPs developed in this work and the commercial SiP, which acts as a reference for benchmarking. Both LTCC SiPs are able to measure the target distance correctly. The LTCC SiP I outperforms the commercial SiP, since it provides a higher peak-to-peak ADC signal and a stronger target peak in the IF spectrum. An improved performance is achieved because the Tx and Rx array gain of the mm-wave module employed in the LTCC SiP I is higher than the off-chip antenna gain of the commercial SiP. The key selling points of the LTCC SiP I in comparison to the commercial SiP are summarized as follows. First, the off-chip circuitry of the LTCC SiP is manufactured at a lower cost, since moderate values of minimum line and space widths are employed and a standard metallization is used. Second, the multi-step cavity structure of the LTCC SiP is more suitable for a selective molding encapsulation. Third, the LTCC SiP I not only withstands the typical LTCC tolerances, e.g. shrinkage and layer misalignment, but also delivers a higher Tx and Rx antenna gain and hence a higher target peak without extending the overall SiP footprint.

In the near future, the resolution of the LTCC structures is expected to show a tremendous improvement. High-resolution LTCC test structures with narrow line and space width of 30 μm have been already manufactured. In addition,

LTCC test structures with microvias of 50 μm diameter have also been demonstrated. The high-resolution structures combined with the multilayer capabilities, excellent material characteristics and low material and production cost of the LTCC technology hold an enormous potential for realizing compact and robust mm-wave SiPs for a wide variety of mass-market consumer applications.

List of Figures

List of Tables

Bibliography

[AJ84] N. Alexopoulos and D. Jackson. Fundamental Superstrate (Cover) Effects on Printed Circuit Antennas. *IEEE Transactions on Antennas and Propagation*, 32(8):807–816, Aug 1984.

[ALHV10] R. E. Amaya, M. Li, K. Hettak, and C. J. Verver. A Broadband 3D Vertical Microstrip to Stripline Transition in LTCC Using a Quasi-Coaxial Structure for Millimetre-Wave SOP Applications. In *The 40th European Microwave Conference*, pages 109–112, Sept 2010.

[BBR+90] H. E. Bair, D. J. Boyle, J. T. Ryan, C. R. Taylor, S. C. Tighe, and D. L. Crouthamel. Thermomechanical Properties of IC Molding Compounds. *Polymer Engineering and Science*, 30(10):609–617, May 1990.

[Bec09] F. Bechtold. A Comprehensive Overview on Today's Ceramic Substrate Technologies. In *2009 European Microelectronics and Packaging Conference*, Rimini, Italy, Jun 2009.

[Bee13] S. Beer. *Methoden und Techniken zur Integration von 122 GHz Antennen in miniaturisierte Radarsensoren*. phd thesis, Karlsruhe Institute of Technology, Faculty of Electronics and Information Technology, 2013.

[BGY+13] S. Beer, M. G. Girma, S. Yaoming, W. Winkler, W. Debski, J. Paaso, G. Kunkel, J. C. Scheytt, J. Hasch, and T. Zwick. Flip-chip Package With Integrated Antenna on a Polyimide Substrate for a 122 GHz Bistatic Radar IC. In *2013 7th European Conference on Antennas and Propagation (EuCAP)*, pages 121–125, Apr 2013.

[BJ89] J. J. Burke and R. W. Jackson. Surface-to-Surface Transition Via Electromagnetic Coupling of Microstrip and Coplanar Waveguide. *IEEE Transactions on Microwave Theory and Techniques*, 37(3):519–525, Mar 1989.

[BPR+12] S. Beer, L. Pires, C. Rusch, J. Paaso, and T. Zwick. A 122 GHz Microstrip Slot Antenna with Via-Fence Resonator in LTCC Technology. In *2012 6th European Conference on Antennas and Propagation (EUCAP)*, pages 1329–1332, Mar 2012.

[BRG+13] S. Beer, C. Rusch, H. Gulan, B. Goettel, M. G. Girma, J. Hasch, W. Winkler, W. Debski, and T. Zwick. An Integrated 122 GHz Antenna Array With Wire Bond Compensation for SMT Radar Sensors. *IEEE Transactions on Antennas and Propagation*, 61(12):5976–5983, Dec 2013.

[BWMS13] F. Bauer, X. Wang, W. Menzel, and A. Stelzer. A 79 GHz Radar Sensor in LTCC Technology Using Grid Array Antennas. *IEEE Transactions on Microwave Theory and Techniques*, 61(6):2514–2521, Jun 2013.

[BZ10] S. Beer and T. Zwick. Probe Based Radiation Pattern Measurements for Highly Integrated Millimeter-Wave Antennas. In *Proceedings of the Fourth European Conference on Antennas and Propagation*, pages 1–5, Apr 2010.

[CAP11] D. J. Chung, A. L. Amadjikpe, and J. Papapolymerou. Multilayer Integration of Low-Cost 60 GHz Front-End Transceiver on Organic LCP. *IEEE Antennas and Wireless Propagation Letters*, 10:1329–1332, 2011.

[Cho15] J. H. Choi, editor. *High-Speed Devices and Circuits with THz Applications*. CRC Press, 1st edition, 2015.

[CI06] C. Caloz and T. Itoh. *Electromagentic Metamaterials : Transmission Line Theory and Microwave Applications*. John Wiley & Sons, Inc., New Jersey, USA, 2006.

[Eme97] D.T. Emerson. The Work of Jagadis Chandra Bose: 100 Years of mm-Wave Research. In *1997 IEEE MTT-S International Microwave Symposium Digest*, pages 553–556, Denver, CO, USA, Jun 1997.

[FAS18] M. Furqan, F. Ahmed, and A. Stelzer. A SiGe 122 GHz FMCW Radar Sensor with 21.5 dBm EIRP Based on a 43-Element Antenna Array in an eWLB Package. In *2018 IEEE MTT-S International Conference on Microwaves for Intelligent Mobility (ICMIM)*, pages 1–4, Apr 2018.

[FTH+14] A. Fischer, Z. Tong, A. Hamidipour, L. Maurer, and A. Stelzer. 77 GHz Multi-Channel Radar Transceiver With Antenna in Package. *IEEE Transactions on Antennas and Propagation*, 62(3):1386–1394, Mar 2014.

[GBBI01] R. Garg, P. Bhartia, I. Bahl, and A. Ittipiboon. *Microstrip Antenna Design Handbook*. Artech House, 2001.

[GSW+17] N. Gutzeit, A. Schulz, T. Welker, C. Wagner, E. Schaefer, and J. Mueller. High-Precision Picosecond Laser Structuring on LTCC for Silicon Chip Assembly with High Electrical Contact Density. In *2017 21st European Microelectronics and Packaging Conference (EMPC) Exhibition*, pages 1–4, Sept 2017.

[GWB+18] B. Goettel, W. Winkler, A. Bhutani, F. Boes, M. Pauli, and T. Zwick. Packaging Solution for a Millimeter-Wave System-on-Chip Radar. *IEEE Transactions on Components, Packaging and Manufacturing Technology*, 8(1):73 – 81, Jan 2018.

[HBRZ13] C. Heine, S. Beer, C. Rusch, and T. Zwick. Via-fence Antennas on LTCC for Radar Applications at 122 GHz. In *2013 European Microwave Conference*, pages 40–43, Oct 2013.

[HWP+18] A. Hagelauer, M. Wojnowski, K. Pressel, R. Weigel, and D. Kissinger. Integrated Systems-in-Package: Heterogeneous Integration of Millimeter-Wave Active Circuits and Passives in Fan-Out Wafer-Level Packaging Technologies. *IEEE Microwave Magazine*, 19(1):48–56, Jan 2018.

[IDBL07] F. P. Incropera, D. P. Dewitt, T. L. Bergman, and A. S. Lavine. *Fundamentals of Heat and Mass Transfer*. John Wiley & Sons, 6th edition, 2007.

[Ima05] Y. Imanaka. *Multilayered Low Temperature Cofired Ceramic (LTCC) Technology*. Springer, New York, USA, 2005.

[JG93] H. Johnson and M. Graham. *High Speed Digital Design: A Handbook of Black Magic*. Prentice Hall PTR, 1993.

[JG11] H. Johnson and M. Graham. *High-Speed Signal Propagation: Advanced Black Magic*. Prentice Hall PTR, July 2011.

[JM87] R. W. Jackson and D. W. Matolak. Surface-to-Surface Transition via Electromagnetic Coupling of Coplanar Waveguides. *IEEE Transactions on Microwave Theory and Techniques*, 35(11):1027–1032, Nov 1987.

[KBP+11] W. T. Khan, S. Bhattacharya, C. Patterson, G. E. Ponchak, and J. Papapolymerou. Low Cost 60 GHz RF Front End Transceiver Integrated on Organic Substrate. In *2011 IEEE MTT-S International Microwave Symposium*, pages 1–4, Jun 2011.

[Kil88] P. S. Kildal. Definition of Artificially Soft and Hard Surfaces for Electromagnetic Waves. *Electronics Letters*, 24(3):168–170, Feb 1988.

[Kil90] P. S. Kildal. Artificially Soft and Hard Surfaces in Electromagnetics. *IEEE Transactions on Antennas and Propagation*, 38(10):1537–1544, Oct 1990.

[KK03] P. S. Kildal and A. Kishk. EM Modeling of Surface with Stop and Go Characteristics - Artificial Magnetic Conductors and Soft and Hard Surfaces. *Applied Computational Electromagnetics Society Journal*, 18(1):32 – 39, Mar 2003.

[KKM05] P. S. Kildal, A. A. Kishk, and S. Maci. Special Issue on Artificial Magnetic Conductors, Soft/Hard Surfaces, and Other Complex Surfaces. *IEEE Transactions on Antennas and Propagation*, 53(1):2–7, Jan 2005.

[KLN+11] D. G. Kam, D. Liu, A. Natarajan, S. Reynolds, H. Chen, and B. A. Floyd. LTCC Packages With Embedded Phased-Array Antennas for 60 GHz Communications. *IEEE Microwave and Wireless Components Letters*, 21(3):142–144, Mar 2011.

[KU12] P. Kovacs and T. Urbanec. Electromagnetic Band Gap Structures: Practical Tips and Advice for Antenna Engineers. *Radioengineering*, 21:414–421, Apr 2012.

[LCP+16] S. Li, T. Chi, J. S. Park, W. T. Khan, H. Wang, and J. Papapolymerou. A Fully Packaged D-band MIMO Transmitter Using High-Density Flip-Chip Interconnects on LCP Substrate. In *2016 IEEE MTT-S International Microwave Symposium (IMS)*, pages 1–4, May 2016.

[LDT+03] R. L. Li, G. DeJean, M. M. Tentzeris, J. Laskar, and J. Papapolymerou. LTCC Multilayer Based CP Patch Antenna Surrounded by a Soft-and-Hard Surface for GPS Applications. In *IEEE Antennas and Propagation Society International Symposium. Digest. Held in conjunction with: USNC/CNC/URSI North Ameri-*

can Radio Sci. Meeting (Cat. No.03CH37450), volume 2, pages 651–654 vol.2, Jun 2003.

[Lee08] Y. C. Lee. CPW-to-Stripline Vertical via Transitions for 60 GHz LTCC SoP Applications. *Progress in Electromagnetics Research Letters*, 2:37 – 44, 2008.

[Lew60] L. Lewin. Radiation from Discontinuities in Strip-Line. *Proceedings of the IEE - Part C: Monographs*, 107(12):163–170, Sept 1960.

[LGO05] S. Lei, Y. X. Guo, and L. C. Ong. CPW to Stripline Transitions in LTCC for Millimeter-Wave Applications. In *2005 Asia-Pacific Microwave Conference Proceedings*, volume 2, pages 3 pp.–, Dec 2005.

[LLI06] C. J. Lee, K. M. K. H. Leong, and T. Itoh. Composite Right/Left-Handed Transmission Line Based Compact Resonant Antennas for RF Module Integration. *IEEE Transactions on Antennas and Propagation*, 54(8):2283–2291, Aug 2006.

[Low18] I. M. Low. *Advances in Ceramic Matrix Composites : Second Edition*. Woodhead Publishing, 2nd edition, 2018.

[LS11] A. Lamminen and J. Saeily. Wideband Stacked Patch Antenna Array on LTCC for W-band. In *Proceedings of the 5th European Conference on Antennas and Propagation (EUCAP)*, pages 2962–2966, Apr 2011.

[Mal12] K. Malecha. Fabrication of Cavities in Low Loss LTCC Materials for Microwave Applications. *Journal of Micromechanics and Microengineering*, 22(12):1–9, Oct 2012.

[MPT+06] J. Mueller, R. Perrone, H. Thust, K. . Drue, C. Kutscher, R. Stephan, J. Trabert, M. Hein, D. Schwanke, J. Pohlner, G. Reppe, R. Kulke, P. Uhlig, A. F. Jacob, T. Baras, and

A. Molke. Technology Benchmarking of High Resolution Structures on LTCC for Microwave Circuits. In *2006 1st Electronic Systemintegration Technology Conference*, volume 1, pages 111–117, Sept 2006.

[NMT+13] D. M. Nair, W. E. McKinzie, B. A. Thrasher, M. A. Smith, E. D. Hughes, and J. M. Parisi. A 10 MHz to 100 GHz LTCC CPW-to-Stripline Vertical Transition. In *2013 IEEE MTT-S International Microwave Symposium Digest (MTT)*, Seattle, WA, USA, Jun 2013.

[SAA+17] S. Scherr, R. Afroz, S. Ayhan, S. Thomas, T. Jaeschke, S. Marahrens, A. Bhutani, M. Pauli, N. Pohl, and T. Zwick. Influence of Radar Targets on the Accuracy of FMCW Radar Distance Measurements. *IEEE Transactions on Microwave Theory and Techniques*, 65(10):3640 – 3647, Oct 2017.

[Sam18] K. K. Samanta. Ceramics for the Future: Advanced Millimeter-Wave Multilayer Multichip Module Integration and Packaging. *IEEE Microwave Magazine*, 19(1):22 – 35, Jan 2018.

[SCF+12] J. Schindelin, I. A. Carreras, E. Frise, V. Kaynig, M. Longair, T. Pietzsch, S. Preibisch, C. Rueden, S. Saalfeld, B. Schmid, J. Y. Tinevez, D. J. White, V. Hartenstein, K. Eliceiri, P. Tomancak, and A. Cardona. Fiji: An Open-Source Platform for Biological-Image Analysis. *Nature Methods*, 9(7):676 – 682, Jun 2012.

[Sie99] D. F. Sievenpiper. *High-Impedance Electromagnetic Surfaces*. PhD thesis, University of California, Los Angeles, 1999.

[Sim01] R. N. Simons. *Coplanar Waveguide Circuits, Components, and Systems*. John Wiley & Sons, 2001.

[SSLL02] D. Sievenpiper, J. Schaffner, J. J. Lee, and S. Livingston. A Steerable Leaky-Wave Antenna Using a Tunable Impedance Ground

Plane. *IEEE Antennas and Wireless Propagation Letters*, 1:179–182, 2002.

[SZB+99] D. Sievenpiper, L. Zhang, R.F.J. Broas, N.G. Alexopolous, and E. Yablonovitch. High Impedance Electromagnetic Surfaces with a Forbidden Frequency Band. *IEEE Transactions on Microwave Theory and Techniques*, 47(11):2059 – 2074, Nov 1999.

[SZC+08] M. Sun, Y. P. Zhang, K. M. Chua, L. L. Wai, D. Liu, and B. P. Gaucher. Integration of Yagi Antenna in LTCC Package for Differential 60 GHz Radio. *IEEE Transactions on Antennas and Propagation*, 56(8):2780–2783, Aug 2008.

[TRK99] R. Tummala, E. J. Rymaszewski, and A. G. Klopfenstein. *Microelectronics Packaging Handbook*, volume 2. Kluwer Academic Publishers, Massachusetts, USA, 2nd edition, 1999.

[TTH02] T. Thelemann, H. Thust, and M. Hintz. Using LTCC for Microsystems. *Microelectronics International*, 19(3):19–23, 2002.

[Tum01] R. R. Tummala. *Fundamentals of Microsystems Packaging*. McGraw-Hill, New York, 1st edition, 2001.

[TWP98] S. D. Targonski, R. B. Waterhouse, and D. M. Pozar. Design of Wide-Band Aperture-Stacked Patch Microstrip Antennas. *IEEE Transactions on Antennas and Propagation*, 46(9):1245–1251, Sept 1998.

[Wen05] J. Wenger. Automotive Radar - Status and Perspectives. In *IEEE Compound Semiconductor Integrated Circuit Symposium, 2005. CSIC '05*, pages 21–24, Palm Springs, CA, USA, Nov 2005.

[Wol06] I. Wolff. *Coplanar Microwave Integrated Circuits*. John Wiley & Sons, 2006.

[WWL+12] M. Wojnowski, C. Wagner, R. Lachner, J. Boeck, G. Sommer, and K. Pressel. A 77 GHz SiGe Single-Chip Four-Channel

Transceiver Module with Integrated Antennas in Embedded Wafer-Level BGA Package. In *2012 IEEE 62nd Electronic Components and Technology Conference*, pages 1027–1032, May 2012.

[YS09] F. Yang and Y. R. Samii. *Electromagentic Band Gap Structures in Antenna Engineering*. Cambridge University Press, New York, USA, 2009.

[ZB12] T. Zwick and S. Beer. QFN Based Packaging Concepts for Millimeter-Wave Transceivers. In *2012 IEEE International Workshop on Antenna Technology (iWAT)*, pages 335–338, Mar 2012.

[ZGZ+15] B. Zhang, H. Gulan, T. Zwick, Y. Li, U. Oderfaelt, F. Carlsson, and H. Zirath. Integration of a 140 GHz Packaged LTCC Grid Array Antenna With an InP Detector. *IEEE Transactions on Components, Packaging and Manufacturing Technology*, 5(8):1060–1068, Aug 2015.

[ZL09] Y. P. Zhang and D. Liu. Antenna-on-Chip and Antenna-in-Package Solutions to Highly Integrated Millimeter-Wave Devices for Wireless Communications. *IEEE Transactions on Antennas and Propagation*, 57(10):2830 – 2841, Oct 2009.

[ZM04] L. Zhu and W. Menzel. Broad-Band Microstrip-to-CPW Transition Via Frequency-Dependent Electromagnetic Coupling. *IEEE Transactions on Microwave Theory and Techniques*, 52(5):1517–1522, May 2004.

Online sources

[Ass16] TJ Green Associates. Hermetic vs Near Hermetic Packaging: A Technical Review. https://www.tjgreenllc.com/2016/09/21/hermetic-vs-near-hermetic-packaging-a-technical-review/, Sept 2016. [Online; accessed 11.Oct.2018].

[Asy] Asymtek. Dam and Fill Encapsulation for Microelectronic Packages. https://smtnet.com/library/files/upload/Dam-and-Fill-Encapsulation.pdf. [Online; accessed 11.Oct.2018].

[Cer] AdTech Ceramics. Design Guidelines: Material and Process Selection. http://www.adtechceramics.com/pdf/AdTech-Ceramics-Design-Guide-Capabilities.pdf. [Online; accessed 11.Oct.2018].

[Com18] CompuPhase. Reflow Soldering Profiles. https://www.compuphase.com/electronics/reflowsolderprofiles.htm, Mar 2018. [Online; accessed 11.Oct.2018].

[Coo] Inc. Gaiser products group CoorsTek. Gaiser Precision Bonding Tools: Wedge Bonding. http://www.nordsonet.com/~/media/Files/Nordson/dage/Tools-Materials/Coors-Tek/Bond-Wedges.pdf. [Online; accessed 11.Oct.2018].

[Cor15a] Ferro Corporation. Technical Data Sheet: A6M/A6M-E High Frequency LTCC Tape System. https://www.

etsmtl.ca/Unites-de-recherche/LTCC/Services-offerts/%Ferro_A6M.pdf, Nov 2015. [Online; accessed 11.Oct.2018].

[Cor15b] Rogers Corporation. Rogers High-Temperature LCP Laminates Ease the Construction of High-Frequency, High-Speed Multilayer Circuit Boards. http://www.rogerscorp.com/news/422/rogers%E2%80%99-high-temperature-lcp-laminates-ease-the-construction-of-high-frequency-high-speed-multilayer-circuit-boards.aspx, Feb 2015. [Online; accessed 11.Oct.2018].

[Elea] Heraeus Electronics. Technical Data Sheet: LTCC Materials - HL2000. https://www.heraeus.com/media/media/het/doc_het/products_and_solutions_het_documents/thick_film/thick_film_data_sheets/passive_components/ltcc/LTCC_Materials_HL2000.pdf. [Online; accessed 11.Oct.2018].

[Eleb] ESL Electroscience. Technical Data Sheet: Co-fire Ceramic Tape 41020. https://www.telecom-bretagne.eu/data/LTCC/ESL41020.pdf. [Online; accessed 11.Oct.2018].

[FaKIoT] Test Facility and Synchrotron Radiation Source at Karlsruhe Institute of Technology. Hard X-ray Imaging - Imaging of Micro Structures and Spatial Inhomogeneities Within Object. https://www.anka.kit.edu/1395.php. [Online; accessed 11.Oct.2018].

[GHM+15] Mart Graef, Bert Huizing, Reinhard Mahnkopf, Hidemi Ishiuchi, Yoshihiro Hayashi, Nobuyuki Ikumi, Hiroyuki Miyakawa, Siyoung Choi, Jae Hoon Choi, Wilman Tsai Sam Pam, Paolo Gargini, Taffy Kingscott, and Linda Wilson. International Technology Roadmap for Semiconductors 2.0. https://ww

w.semiconductors.org/clientuploads/Research_Technology/ITRS/2015/4_2015%20ITRS%202.0_OSC%20.pdf, 2015. [Online; accessed 11-October-2018].

[Gmb] Micro Hybrid Electronic GmbH. LTCC Multi Layer Ceramics. https://www.micro-hybrid.de/en/products/electronic-micro-systems/ltcc-multilayerceramic.html. [Online; accessed 11.Oct.2018].

[Gmb12] Polytec PT GmbH. Polytec Adhesives Selector Guide. https://www.polytec-pt.com/int/products/epoxy-adhesives/, Mar 2012. [Online; accessed 11.Oct.2018].

[Gmb15] Polyec PT GmbH. Data Sheet Polytec TC 430. http://old.polytec.com/fileadmin/user_uploads/Polytec/Polytec_France/Produits/documents/Polytec_TC_430_engl_v2015.pdf, 2015. [Online; accessed 11.Oct.2018].

[Gmb18] Silicon Radar GmbH. 120 GHz Highly Integrated IQ Transceiver with Antennas in Package in Silicon Germanium Technology. https://www.siliconradar.com/datasheets/Datasheet_TRX_120_001_V1.3.pdf, Sept 2018. [Online; accessed 11.Oct.2018].

[GO15] DuPont Microcircuit Materials Glenn Oliver. Low-loss Materials in High Frequency Electronics and The Challenges of Measurement. https://wiki.epfl.ch/thz/documents/Materials%20and%20Measurements%20-%20DuPont%20-%20Feb2015.pdf, Feb 2015. [Online; accessed 11.Oct.2018].

[IKT] Fraunhofer IKTS. LTCC Tapes According To Heraeus Formula. https://www.ikts.fraunhofer.de/content/dam/ikts/downloads/electronics_and_microsystems/hybrid_microsystems/IKTS_326_D_LTCC-Folien%20nach%20He raeus-Rezeptur_final_2016-04-16_ENx.pdf. [Online; accessed 11.Oct.2018].

[Inc14] Master Bond Inc. Protecting Chip on Board (COB) Devices with Glob Top Encapsulants. https://www.electronicproducts.com/Packaging_and_Hardware/Adhesives_Sealants_Coatings/Protecting_Chip_on_Board_COB_Devices_with_Glob_Top_Encapsulants.aspx, Mar 2014. [Online; accessed 11.Oct.2018].

[Kyo] Kyocera. Ceramic Packages: Material Properties. https://global.kyocera.com/prdct/semicon/material/. [Online; accessed 11.Oct.2018].

[Ltd] Plextek RFI Ltd. Plextek RF Integration: Recommendations and Advice for the Successful Design of LTCC Components. https://www.plextekrfi.com/wp-content/uploads/LTCC_technology_overview.pdf. [Online; accessed 11.Oct.2018].

[Mat09] DuPont Microcircuit Materials. DuPont GreenTape Low Temperature Co-fired Ceramic System: Design and Layout Guidelines. http://www.dupont.com/content/dam/dupont/products-and-services/electronic-and-electrical-materials/documents/prodlib/GreenTape_Design_Layout_Guidelines.pdf, Nov 2009. [Online; accessed 11.Oct.2018].

[Mat18] DuPont Microcircuit Materials. GreenTape 9K7 LTCC Material System. http://www.dupont.com/products-and-services/electronic-electrical-materials/low-temperature-co-fire-ceramic-materials/brands/greentape-ceramic-circuit/products/green-tape-9k7-ltcc-material-system.html, 2018. [Online; accessed 11.Oct.2018].

[Top14] Topsil. High Resistivity (HiRes) Silicon for GHz and THz Technology. http://www.topsil.com/media/123119/hires_application_note_v1.1_january2014.pdf, January 2014. [Online; accessed 11.Oct.2018].

[uSGCK] Wellenzahl Radar und Sensortechnik GmbH & Co. KG. 122 GHz Radar Sensor RAPID v3. https://wellenzahl.de/index.php?id=5. [Online; accessed 11.Oct.2018].

[YG18] Chip Yorkgitis and Joshua Guyan. Beyond the Latest Frontier: Licensed, Unlicensed and Experimental Operations above 95 GHz. https://www.commlawmonitor.com/2018/02/articles/wireless-2/beyond-the-latest-frontier-licensed-unlicensed-and-experimental-operations-above-95-ghz/, Feb 2018. [Online; accessed 11.Oct.2018].

Journal Publications

[1] S. Scherr, S. Ayhan, B. Fischbach, A. Bhutani, M. Pauli, and T. Zwick. An Efficient Frequency and Phase Estimation Algorithm With CRB Performance for FMCW Radar Applications. *IEEE Transactions on Instrumentation and Measurement*, 64(7):1868–1875, Jul 2015.

[2] S. Ayhan, S. Scherr, A. Bhutani, B. Fischbach, M. Pauli, and T. Zwick. Impact of Frequency Ramp Nonlinearity, Phase Noise, and SNR on FMCW Radar Accuracy. *IEEE Transactions on Microwave Theory and Techniques*, 64(10):3290–3301, Oct 2016.

[3] S. Ayhan, M. Pauli, S. Scherr, B. Goettel, A. Bhutani, S. Thomas, T. Jaeschke, J. Panel, F. Vivier, L. Eymard, A. Weill, N. Pohl, and T. Zwick. Millimeter-Wave Radar Sensor for Snow Height Measurements. *IEEE Transactions on Geoscience and Remote Sensing*, 55(2):854–861, Feb 2017.

[4] A. Bhutani, B. Goettel, T. Thelemann, and T. Zwick. CPW-to-SL Transition in LTCC Technology. *Electronics Letters*, 53(9):609–611, 2017.

[5] M. Pauli, B. Goettel, S. Scherr, A. Bhutani, S. Ayhan, W. Winkler, and T. Zwick. Miniaturized Millimeter-Wave Radar Sensor for High-Accuracy Applications. *IEEE Transactions on Microwave Theory and Techniques*, 65(5):1707–1715, May 2017.

[6] S. Scherr, R. Afroz, S. Ayhan, S. Thomas, T. Jaeschke, S. Marahrens, A. Bhutani, M. Pauli, N. Pohl, and T. Zwick. Influence of Radar Targets

on the Accuracy of FMCW Radar Distance Measurements. *IEEE Transactions on Microwave Theory and Techniques*, 65(10):3640–3647, Oct 2017.

[7] T. Zwick, F. Boes, B. Goettel, A. Bhutani, and M. Pauli. Pea-Sized mmW Transceivers: QFN-Based Packaging Concepts for Millimeter-Wave Transceivers. *IEEE Microwave Magazine*, 18(6):79–89, Sept 2017.

[8] B. Goettel, W. Winkler, A. Bhutani, F. Boes, M. Pauli, and T. Zwick. Packaging Solution for a Millimeter-Wave System-on-Chip Radar. *IEEE Transactions on Components, Packaging and Manufacturing Technology*, 8(1):73–81, Jan 2018.

[9] D. Mueller, A. Haag, A. Bhutani, A. Tessmann, A. Leuther, T. Zwick, and I. Kallfass. Bandwidth Optimization Method for Reflective-Type Phase Shifters. *IEEE Transactions on Microwave Theory and Techniques*, 66(4):1754–1763, Apr 2018.

[10] A. Bhutani, B. Goettel, A. Lipp, and T. Zwick. Packaging Solution Based on Low-Temperature Cofired Ceramic Technology for Frequencies Beyond 100 GHz. *IEEE Transactions on Components, Packaging and Manufacturing Technology*, 9(5):945–954, May 2019.

Conference Publications

[11] S. Ayhan, S. Scherr, A. Bhutani, M. Pauli, and T. Zwick. Radar Sensor for Waveguide Based Distance Measurements in Machine Tool Components. In *2015 IEEE MTT-S International Conference on Microwaves for Intelligent Mobility (ICMIM)*, pages 1–4, Apr 2015.

[12] A. Bhutani, B. Goettel, D. Mueller, and T. Zwick. Novel Planar Electromagnetic Bandgap for Mutual Coupling Reduction Between Patch Antennas. In *2015 German Microwave Conference*, pages 76–79, Mar 2015.

[13] B. Goettel, H. Gulan, A. Bhutani, M. Pauli, and T. Zwick. Ultra Broadband Multiple Feed Antenna for Efficient On-Chip Power Combining. In *2015 IEEE International Symposium on Antennas and Propagation USNC/URSI National Radio Science Meeting*, pages 2037–2038, Jul 2015.

[14] S. Scherr, B. Goettel, S. Ayhan, A. Bhutani, M. Pauli, W. Winkler, J. C. Scheytt, and T. Zwick. Miniaturized 122 GHz ISM Band FMCW Radar With Micrometer Accuracy. In *2015 European Radar Conference (EuRAD)*, pages 277–280, Sept 2015.

[15] S. Ayhan, S. Scherr, A. Bhutani, M. Pauli, and T. Zwick. System Simulation for FMCW Radar in Industrial Applications. In *2016 IEEE Topical Conference on Wireless Sensors and Sensor Networks (WiSNet)*, pages 27–30, Jan 2016.

[16] A. Bhutani, B. Goettel, T. Streitz, S. Scherr, W. Winkler, and T. Zwick. Low-Cost Antenna-in-Package Solution for 122 GHz Radar Module. In *2016 46th European Microwave Conference (EuMC)*, pages 1469–1472, Oct 2016.

[17] A. Bhutani, H. Gulan, B. Goettel, C. Heine, T. Thelemann, M. Pauli, and T. Zwick. 122 GHz Aperture-Coupled Stacked Patch Microstrip Antenna in LTCC Technology. In *2016 10th European Conference on Antennas and Propagation (EuCAP)*, pages 1–5, Apr 2016.

[18] A. Bhutani, J. Schaefer, M. Pauli, S. Scherr, B. Goettel, M. Nierlich, and T. Zwick. 3D Metal Printed Ku/Ka Band Modified Turnstile Junction Orthomode Transducer. In *2016 Asia-Pacific Microwave Conference (APMC)*, pages 1–4, Dec 2016.

[19] B. Goettel, D. Mueller, H. Gulan, A. Bhutani, and T. Zwick. Active Multiple-Feed on-Chip Radiator With in-Antenna Power-Combining Approach. In *2016 10th European Conference on Antennas and Propagation (EuCAP)*, pages 1–5, April 2016.

[20] B. Goettel, J. Schaefer, A. Bhutani, H. Gulan, and T. Zwick. In-Antenna Power-Combining Methods. In *2017 11th European Conference on Antennas and Propagation (EUCAP)*, pages 2776–2730, Mar 2017.

[21] A. Bhutani, B. Goettel, J. Mayer, M. Pauli, and T. Zwick. Electromagnetically Coupled Coplanar Waveguide to Stripline Transition in LTCC Technology. In *2018 11th German Microwave Conference (GeMiC)*, pages 371–374, Mar 2018.

[22] A. Bhutani, B. Goettel, J. Schaefer, and T. Zwick. Partially Molded Antenna-in-Package Concept for 122 GHz SiGe Radar Sensor. In *12th European Conference on Antennas and Propagation (EuCAP 2018)*, pages 1–4, London, UK, Apr 2018.

[23] A. Bhutani, B. Goettel, J. Mayer, M. Pauli, and T. Zwick. 122 GHz Aperture-Coupled Mushroom EBG Antenna in LTCC Technology. In

13th European Conference on Antennas and Propagation (EuCAP), pages 1–5, Krakow, Poland, April 2019.

[24] A. Bhutani, B. Goettel, M. Pauli, and T. Zwick. 122 GHz FMCW Radar System-in-Package in LTCC Technology. In *49th European Microwave Conference (EuMC)*, pages 1–4, Paris, October 2019. Accepted for publication.

Forschungsberichte aus dem Institut für Höchstfrequenztechnik und Elektronik (IHE) der Universität Karlsruhe (TH)
(ISSN 0942-2935)

Herausgeber: Prof. Dr.-Ing. Dr. h.c. Dr.-Ing. E.h. mult. Werner Wiesbeck

Die Bände 1 (1992) bis 55 (2008) der Schriftenreihe können über das Institut Hochfrequenztechnik und Elektronik bestellt werden (https://www.ihe.kit.edu).

Fortführung als:

Karlsruher Forschungsberichte aus dem Institut für Hochfrequenztechnik und Elektronik
(ISSN 1868-4696)

Herausgeber: Prof. Dr.-Ing. Thomas Zwick

Band 55	Sandra Knörzer **Funkkanalmodellierung für OFDM-Kommunikationssysteme bei Hochgeschwindigkeitszügen (2009)** ISBN 978-3-86644-361-7
Band 56	Thomas Fügen **Richtungsaufgelöste Kanalmodellierung und Systemstudien für Mehrantennensysteme in urbanen Gebieten (2009)** ISBN 978-3-86644-420-1
Band 57	Elena Pancera **Strategies for Time Domain Characterization of UWB Components and Systems (2009)** ISBN 978-3-86644-417-1
Band 58	Jens Timmermann **Systemanalyse und Optimierung der Ultrabreitband-Übertragung (2010)** ISBN 978-3-86644-460-7

Die Bände sind unter www.ksp.kit.edu als PDF frei verfügbar oder als Druckausgabe bestellbar.

Band 59 Juan Pontes
Analysis and Design of Multiple Element Antennas for Urban Communication (2010)
ISBN 978-3-86644-513-0

Band 60 Andreas Lambrecht
True-Time-Delay Beamforming für ultrabreitbandige Systeme hoher Leistung (2010)
ISBN 978-3-86644-522-2

Band 61 Grzegorz Adamiuk
Methoden zur Realisierung von dual-orthogonal, linear polarisierten Antennen für die UWB-Technik (2010)
ISBN 978-3-86644-573-4

Band 62 Jutta Kühn
AlGaN/GaN-HEMT Power Amplifiers with Optimized Power-Added Efficiency for X-Band Applications (2011)
ISBN 978-3-86644-615-1

Band 63 Małgorzata Janson
Hybride Funkkanalmodellierung für ultrabreitbandige MIMO-Systeme (2011)
ISBN 978-3-86644-639-7

Band 64 Mario Pauli
Dekontaminierung verseuchter Böden durch Mikrowellenheizung (2011)
ISBN 978-3-86644-696-0

Band 65 Thorsten Kayser
Feldtheoretische Modellierung der Materialprozessierung mit Mikrowellen im Durchlaufbetrieb (2011)
ISBN 978-3-86644-719-6

Band 66 Christian Andreas Sturm
Gemeinsame Realisierung von Radar-Sensorik und Funkkommunikation mit OFDM-Signalen (2012)
ISBN 978-3-86644-879-7

Band 67 Huaming Wu
Motion Compensation for Near-Range Synthetic Aperture Radar Applications (2012)
ISBN 978-3-86644-906-0

Band 68 Friederike Brendel
Millimeter-Wave Radio-over-Fiber Links based on Mode-Locked Laser Diodes (2013)
ISBN 978-3-86644-986-2

Band 69 Lars Reichardt
Methodik für den Entwurf von kapazitätsoptimierten Mehrantennensystemen am Fahrzeug (2013)
ISBN 978-3-7315-0047-6

Band 70 Stefan Beer
Methoden und Techniken zur Integration von 122 GHz Antennen in miniaturisierte Radarsensoren (2013)
ISBN 978-3-7315-0051-3

Band 71 Łukasz Zwirełło
Realization Limits of Impulse-Radio UWB Indoor Localization Systems (2013)
ISBN 978-3-7315-0114-5

Band 72 Xuyang Li
Body Matched Antennas for Microwave Medical Applications (2014)
ISBN 978-3-7315-0147-3

Band 73 Sebastian Diebold
Transistor- und Leitungsmodellierung zum Entwurf von monolithisch integrierten Leistungsverstärkern für den hohen Millimeterwellen-Frequenzbereich (2014)
ISBN 978-3-7315-0161-9

Band 74 Christian Rusch
Integrierte, planare Leckwellenantennen für 3D-Millimeterwellen-Radarsysteme basierend auf dem holografischen Prinzip (2014)
ISBN 978-3-7315-0234-0

Band 75 Marlene Harter
Dreidimensional bildgebendes Radarsystem mit digitaler Strahlformung für industrielle Anwendungen (2014)
ISBN 978-3-7315-0249-4

Band 76 Michael A. Baldauf
Abhängigkeit der Exposition von der Zellgröße beim Mobilfunk unter Gewährleistung der Versorgung (2015)
ISBN 978-3-7315-0308-8

Band 77 Alicja Ossowska
Highly Resolved Synthetic Aperture Radar with Beam Steering (2015)
ISBN 978-3-7315-0315-6

Band 78 Małgorzata Dominika Brzeska
RF Modelling and Characterization of Tyre Pressure Sensors and Vehicle Access Systems (2015)
ISBN 978-3-7315-0348-4

Band 79 Ulrich Lewark
Aktive Frequenzvervielfacher zur Signalerzeugung im Millimeter- und Submillimeterwellen Frequenzbereich (2015)
ISBN 978-3-7315-0354-5

Band 80 Kai-Philipp Walter Pahl
Distributed Transformers for Broadband Monolithic Millimeter-Wave Integrated Power Amplifiers (2015)
ISBN 978-3-7315-0409-2

Band 81 Serdal Ayhan
Hochgenaue radarbasierte Abstandsmessung mit geführter Wellenausbreitung (2016)
ISBN 978-3-7315-0433-7

Band 82 Yoke Leen Sit
MIMO OFDM Radar-Communication System with Mutual Interference Cancellation (2017)
ISBN 978-3-7315-0599-0

Band 83 Steffen Scherr
FMCW-Radarsignalverarbeitung zur Entfernungsmessung mit hoher Genauigkeit (2017)
ISBN 978-3-7315-0607-2

Band 84 Tom Schipper
Modellbasierte Analyse des Interferenzverhaltens von Kfz-Radaren (2017)
ISBN 978-3-7315-0639-3

Band 85 Malyhe Jalilvand
Application-Specific Broadband Antennas for Microwave Medical Imaging (2017)
ISBN 978-3-7315-0664-5

Band 86 Benjamin Göttel
Millimeterwellen On-Chip Antennensysteme für die Integration in SoC Applikationen (2017)
ISBN 978-3-7315-0667-6

Band 87 Christian Arnold
Im Orbit einstellbare Ausgangsfilter und -multiplexer (2017)
ISBN 978-3-7315-0722-2

Band 88 Tobias Mahler
Synthese kapazitätsoptimierter Antennensysteme mit messtechnischer Verifikation (2018)
ISBN 978-3-7315-0737-6

Band 89 Daniel Müller
RF Probe-Induced On-Wafer Measurement Errors in the Millimeter-Wave Frequency Range (2018)
ISBN 978-3-7315-0822-9

Band 90 Tristan Visentin
Polarimetric Radar for Automotive Applications (2019)
ISBN 978-3-7315-0888-5

Band 91 Christian von Vangerow
Entwurf und Modellierung von Breitbandverstärkern mit variablem Gewinn in SiGe BiCMOS Technologien (2019)
ISBN 978-3-7315-0910-3

Band 92 Mekdes Girma
Concepts for Short Range Millimeter-wave Miniaturized Radar Systems with Built-in Self-Test (2019)
ISBN 978-3-7315-0938-7

Band 93 Akanksha Bhutani
Low Temperature Co-fired Ceramics for System-in-Package Applications at 122 GHz (2019)
ISBN 978-3-7315-0945-5